Meaning in Mathematics Education

Springer

Mathematics Education Library

VOLUME 37

Managing Editor

A.J. Bishop, *Monash University, Melbourne, Australia*

The titles published in this series are listed at the end of this volume.

Jeremy Kilpatrick
Celia Hoyles
Ole Skovsmose
(Editors, in collaboration with Paola Valero)

Meaning in Mathematics Education

Springer

Jeremy Kilpatrick, University of Georgia, Georgia, USA
Celia Hoyles, University of London, London, United Kingdom
Ole Skovsmose, Aalborg University, Aalborg, Denmark
Paola Valero, Aalborg University, Aalborg, Denmark

Library of Congress Cataloging-in-Publication Data

A C.I.P. Catalogue record for this book is available
from the Library of Congress.

ISBN 0-387-24039-X e-ISBN 0-387-24040-3 Printed on acid-free paper.

Printed in the United States of America.

9 8 7 6 5 4 3 2 1 SPIN 11323051

springeronline.com

CONTENTS

PREFACE

This book is a product of the BACOMET group, a group of educators—mainly educators of prospective teachers of mathematics—who first came together in 1980 to engage in study, discussion, and mutual reflection on issues in mathematics education. BACOMET is an acronym for *BA*sic *CO*mponents of *M*athematics *E*ducation for *T*eachers. The group was formed after a series of meetings in 1978–1979 between Geoffrey Howson, Michael Otte, and the late Bent Christiansen. In the ensuing years, BACOMET initiated several projects that resulted in published works. The present book is the main product of the BACOMET project entitled *Meaning and Communication in Mathematics Education*. This theme was chosen because of the growing recognition internationally that teachers of mathematics must deal with questions of meaning, sense making, and communication if their students are to be proficient learners and users of mathematics.

The participants in this project were the following:

Nicolas Balacheff (Grenoble, France)
Maria Bartolini Bussi (Modena, Italy)
Rolf Biehler (Bielefeld, Germany)
Robert Davis (New Brunswick, NJ, USA)
Willibald Dörfler (Klagenfurt, Austria)
Tommy Dreyfus (Holon, Israel)
Joel Hillel (Montreal, Canada)
Geoffrey Howson (Southampton, England)
Celia Hoyles—*Director* (London, England)
Jeremy Kilpatrick—*Director* (Athens, GA, USA)
Christine Keitel (Berlin, Germany)
Colette Laborde (Grenoble, France)
Michael Otte (Bielefeld, Germany)
Kenneth Ruthven (Cambridge, England)
Anna Sierpinska (Montreal, Canada)
Ole Skovsmose—*Director* (Aalborg, Denmark)

Conversations about directions the project might take began in May 1993 at a NATO Advanced Research Workshop of the previous BACOMET project in

Villard-de-Lans, France. The main themes and topics for papers began to emerge at a meeting in June 1994 in Melle, Germany, and were put in final form at a conference in Athens, Georgia, USA, in September 1995, where participants' papers were critiqued and revised, the structure of the book was created, and interludes were drafted to introduce the sections of the book.

Subsequently, several project members either withdrew because of the press of other responsibilities (Davis & Dörfler), decided to publish their work elsewhere (Balacheff), or decided that their papers no longer represented their current thinking (Ruthven). The chapters in this book were written by the remaining project members.

We would like to thank the Scientific Affairs Division of the North Atlantic Treaty Organisation for funding the 1993 workshop and the Office of the Vice President for Academic Affairs of the University of Georgia for funding the 1995 conference. We are grateful to Brian Lawler of the University of Georgia for helping to polish the English of the text. The book would never have seen its way into print without the collaboration of Paola Valero of Aalborg University. Her many contributions were invaluable, and we acknowledge them with pleasure and gratitude.

Bob Davis, charter member of BACOMET, founding editor of the *Journal of Mathematical Behavior,* and a pioneer in studying mathematics education from the perspective of cognitive science, died in 1997. This book is dedicated to his memory.

Jeremy Kilpatrick
Celia Hoyles
Ole Skovsmose

Mathematics Education Library

Managing Editor: A.J. Bishop, Melbourne, Australia

1. H. Freudenthal: Didactical Phenomenology of Mathematical Structures. 1983
ISBN 90-277-1535-1 HB; 90-277-2261-7 PB
2. B. Christiansen, A. G. Howson and M. Otte (eds.): Perspectives on Mathematics Education. Papers submitted by Members of the Bacomet Group. 1986
ISBN 90-277-1929-2 HB; 90-277-2118-1 PB
3. A. Treffers: Three Dimensions. A Model of Goal and Theory Description in Mathematics Instruction TheWiskobas Project. 1987 ISBN 90-277-2165-3
4. S. Mellin-Olsen: The Politics of Mathematics Education. 1987
ISBN 90-277-2350-8
5. E. Fischbein: Intuition in Science and Mathematics. An Educational Approach. 1987 ISBN 90-277-2506-3
6. A.J. Bishop: Mathematical Enculturation. A Cultural Perspective on Mathematics Education.1988 ISBN 90-277-2646-9 HB; 1991 0-7923-1270-8 PB
7. E. von Glasersfeld (ed.): Radical Constructivism in Mathematics Education. 1991 ISBN 0-7923-1257-0
8. L. Streefland: Fractions in Realistic Mathematics Education. A Paradigm of Developmental Research. 1991 ISBN 0-7923-1282-1
9. H. Freudenthal: Revisiting Mathematics Education. China Lectures. 1991
ISBN 0-7923-1299-6
10. A.J. Bishop, S. Mellin-Olsen and J. van Dormolen (eds.): Mathematical Knowledge: Its Growth Through Teaching. 1991 ISBN 0-7923-1344-5
11. D. Tall (ed.): Advanced Mathematical Thinking. 1991 ISBN 0-7923-1456-5
12. R. Kapadia and M. Borovcnik (eds.): Chance Encounters: Probability in Education. 1991 ISBN 0-7923-1474-3
13. R. Biehler, R.W. Scholz, R. Sträßer and B. Winkelmann (eds.): Didactics of Mathematics as a Scientific Discipline. 1994 ISBN 0-7923-2613-X
14. S. Lerman (ed.): Cultural Perspectives on the Mathematics Classroom. 1994
ISBN 0-7923-2931-7
15. O. Skovsmose: Towards a Philosophy of Critical Mathematics Education. 1994
ISBN 0-7923-2932-5
16. H. Mansfield, N.A. Pateman and N. Bednarz (eds.): Mathematics for Tomorrow's Young Children. International Perspectives on Curriculum. 1996
ISBN 0-7923-3998-3
17. R. Noss and C. Hoyles:Windows on Mathematical Meanings. Learning Cultures and Computers. 1996 ISBN 0-7923-4073-6 HB; 0-7923-4074-4 PB
18. N. Bednarz, C. Kieran and L. Lee (eds.): Approaches to Algebra. Perspectives for Research and Teaching.1996 ISBN 0-7923-4145-7 HB; 0-7923-4168-6, PB
19. G. Brousseau: Theory of Didactical Situations in Mathematics. Didactique des Mathématiques 1970-1990. Edited and translated by N. Balacheff, M. Cooper, R. Sutherland and V. Warfield. 1997 ISBN 0-7923-4526-6
20. T. Brown: Mathematics Education and Language. Interpreting Hermeneutics and Post-Structuralism. 1997 ISBN 0-7923-4554-1 HB
Second Revised Edition. 2001 ISBN 0-7923-6969-6 PB

Mathematics Education Library

21. D. Coben, J. O'Donoghue and G.E. FitzSimons (eds.): Perspectives on Adults Learning Mathematics. Research and Practice. 2000 ISBN 0-7923-6415-5
22. R. Sutherland, T. Rojano, A. Bell and R. Lins (eds.): Perspectives on School Algebra. 2000 ISBN 0-7923-6462-7
23. J.-L. Dorier (ed.): On the Teaching of Linear Algebra. 2000 ISBN 0-7923-6539-9
24. A. Bessot and J. Ridgway (eds.): Education for Mathematics in the Workplace. 2000 ISBN 0-7923-6663-8
25. D. Clarke (ed.): Perspectives on Practice and Meaning in Mathematics and Science Classrooms. 2001 ISBN 0-7923-6938-6 HB; 0-7923-6939-4 PB
26. J. Adler: Teaching Mathematics in Multilingual Classrooms. 2001 ISBN 0-7923-7079-1 HB; 0-7923-7080-5 PB
27. G. de Abreu, A.J. Bishop and N.C. Presmeg (eds.): Transitions Between Contexts of Mathematical Practices. 2001 ISBN 0-7923-7185-2
28. G.E. FitzSimons:What Counts as Mathematics? Technologies of Power in Adult and Vocational Education. 2002 ISBN 1-4020-0668-3
29. H.Alrø andO. Skovsmose:Dialogue and Learning in Mathematics Education. Intention, Reflection,Critique.2002 ISBN 1-4020-0998-4 HB;1-4020-1927-0 PB
30. K. Gravemeijer, R. Lehrer, B. van Oers and L. Verschaffel (eds.): Symbolizing, Modeling and Tool Use in Mathematics Education. 2002 ISBN 1-4020-1032-X
31. G.C. Leder, E. Pehkonen and G. Tőrner (eds.): Beliefs: A Hidden Variable in Mathematics Education? 2002 ISBN 1-4020-1057-5 HB; 1-4020-1058-3 PB
32. R. Vithal: In Search of a Pedagogy of Conflict and Dialogue for Mathematics Education. 2003 ISBN 1-4020-1504-6
33. H.W. Heymann: Why Teach Mathematics? A Focus on General Education. 2003 Translated by Thomas LaPresti ISBN 1-4020-1786-3
34. L. Burton: Mathematicians as Enquirers: Learning about Learning Mathematics. 2004 ISBN 1-4020-7853-6 HB; 1-4020-7859-5 PB
35. P. Valero, R. Zevenbergen (eds.): Researching the Socio-Political Dimensions of Mathematics Education: Issues of Power in Theory and Methodology. 2004 ISBN 1-4020-7906-0
36. D. Guin, K. Ruthven, L. Trouche (eds.) The Didactical Challenge of Symbolic Calculators: Turning a Computational Device into a Mathematical Instrument 2005 ISBN 0-387-23158-7
37. J. Kilpatrick, C. Hoyles, O. Skovsmose (eds. in collaboration with Paola Valero) Meaning in Mathematics Education. 2005 ISBN 0-387-24039-X

INTRODUCTION

In philosophy much attention has been paid to the notion of "meaning". Ludwig Wittgenstein introduced *Philosophical Investigations* in 1953 by quoting from Augustine's *Confessions*, which outlines a theory of meaning as a theory of reference, drawing on the Platonic tradition. The theory of meaning, as a refined theory of reference, has, however, developed in many directions.

Gottlob Frege, in *Über Sinn und Bedeutung* [Meaning and Reference] in 1892, made the distinction between *Sinn,* which refers to meaning in a qualitative and experiential way, and *Bedeutung,* signifying the referential aspects of meaning. Frege's work fundamentally influenced later interpretations of logic and mathematics. When, for example, the *Sinn* of "red" is described as the properties of being red, the *Bedeutung* of red can be interpreted as the set of red things. Furthermore, Frege suggested that the *Sinn* of a statement is interpreted as the content of what is claimed by the statement, while the *Bedeutung* of a statement is its truth value, where only two such values exist, namely "true" and "false". A new branch of the philosophy of mathematics was thus established. While logic and the philosophy of mathematics dealt with the meaning of concepts and statements in terms of *Bedeutung,* psychology (and related fields) opened up exploration of meaning in terms of *Sinn*. This division has proved to be a milestone in the philosophy of meaning.

In *Philosophical Investigations*, Wittgenstein presented a different approach to the discussion of meaning. Instead of considering the meaning of concepts and statements in terms of *Sinn* and *Bedeutung,* he suggested that it could be understood in terms of their use. This interpretation broadens the idea of what ascribed meaning can be since many other entities besides concepts and statements have uses. If use signifies meaning, then the scope of theories of meaning could, for example, include gestures, acts, and activities. This leads us towards discourse theory, which demonstrates how discussions of meaning can become a way of reflecting on human activity in a social context.

Even when we turn our attention to a specific field like mathematics, the discussion of meaning maintains its complexity. Historically, the philosophy of mathematics has been part of the philosophical discussion of meaning, and basic steps in the interpretation of meaning have been inspired by a discussion of meaning in mathematics. Thus, Platonism addresses the question: What could be the meaning of a mathematical term like *triangle*? Where is it possible to search for the reference of such a term? Where is it possible to search for the general triangle? And how can these searches be undertaken? According to Platonism, the senses are of no use, as only human reason provides the means to scrutinize the meaning of mathematical concepts, as well as the meaning of other fundamental terms, like *courage, justice,* and *beauty*. Using reason, we can "observe" the world of reference of mathematics, and it is reason that is the human faculty that provides access to the Platonic world.

The discussion of meaning gives rise to different philosophies of mathematics. From classic Platonism, various forms of realism have developed, for example, Frege's set-theoretical realism. When use and not only reference was included as a candidate for meaning, the process of doing mathematics could become central to a "meaning-producing activity". So, with reference to Wittgenstein, Imre Lakatos's *Proofs and Refutations* could be interpreted as a contribution to the discussion of meaning in mathematics. To understand the meaning of a concept, theorem, or mathematical idea, it is important to appreciate the process through which this entity has evolved, which Lakatos described as a dialectic of proofs and refutations. Thus, the constructive processes, rather than the referential elements, provide meaning.

When we consider the question of meaning with respect to mathematics education, the issue becomes even more complex, since philosophical and non-philosophical interpretations of meaning can become mixed. Thus, on the one hand, we may claim that an activity has meaning as part of the curriculum, while students might feel that the same activity is totally devoid of meaning. Observations such as these about meaning can be discussed with reference to empirical data. On the other hand, discussion of whether meaning has to do with reference or use is conceptual and not easily explained by such means. Given this complexity, it is appropriate that this book about meaning in mathematics education encompasses a range of perspectives on the topic and includes discussion of both theoretical and empirical issues.

The book is organised into ten chapters and three interludes. The interludes, entitled "Meanings of *Meaning in Mathematics*", "Collective Meaning and Common Sense" and "Communication and Construction of Meaning", serve to highlight and illustrate general issues in the discussion of meaning with respect to mathematics education: the definitions of meaning, the socio-political dimensions of meaning, and the construction and development of meaning through networks of interaction. All three interludes emerged from discussions in the BACOMET group and were collaboratively written. Since single or co-authored chapters relate more or less explicitly to an interlude, we organise the rest of the introduction by reference to each interlude in turn, followed by chapters that are most closely associated with it.

The interlude "Meanings of *Meaning in Mathematics*" examines some of the ways in which *meaning* is used in mathematics education. For example, when students ask, "What does this mean?" they are often asking why they are doing something or why some statement has been made, so their question involves implicit issues of intention and reference. This general problem of assigning meaning is illustrated with a discussion of decimal numbers, which have one meaning in mathematics, another in commercial practice, and yet another in the classroom. We argue that distinguishing *spheres of practice* in which mathematical concepts take on different meanings is helpful, along with separating meanings *in* school mathematics from meanings given *for* school mathematics.

Various ways of looking at "the meaning of *X*," where *X* is mathematical content to be taught, are examined. We suggest that because meanings in school mathematics appear to be relatively unambiguous, the need for gradual development is often overlooked, with teachers needing to become aware that "everyday" meanings in mathematics are only part of the story.

In "'Meaning' and School Mathematics", Geoffrey Howson bases an exploration of various "meanings" associated with school mathematics on a study of Grade 8 texts drawn from eight countries: England, France, Japan, the Netherlands, Norway, Spain, Switzerland, and the United States. The chapter focuses on three aspects of meaning: (a) What meaning is attached to the study of school mathematics? (b) What do "doing mathematics" and "being a mathematician" mean? and (c) How can one associate "meaning" with mathematical objects and concepts? Thus "relevance" and "What is the point of this for me?" are discussed along with signification and referents.

Howson compares and contrasts the manner in which professional mathematicians attach meaning to concepts with the way this is done in school texts. Special critical attention is paid to the ways in which texts attempt to motivate students and how that interacts with notions of "common sense". Finally, a brief case study considers how different texts attempt to assign meaning to the product of negative numbers, and some inferences are drawn for mathematics education.

In "The Meaning of Conics: Historical and Didactical Dimensions", Mariolina Bartolini-Bussi discusses the different meanings of conics developed during a research project concerning the teaching and learning of geometry in high school. In this project special care was taken to build a context for student activity that offered a pragmatic basis for the knowledge to be learned through historical sources and physical models.

The core thesis developed by Bartolini-Bussi is that the present meaning of *conics* is the result of complex relationships between the different processes of studying conics in different historical periods, each of which has left some residue in terminology, problems, means of representation, rules of actions, and systems of control. As a consequence, it is not possible to build the meaning of conics through just one approach, for instance, through algebraic definition. Also, if history is an unavoidable component of the construction of meaning of conics, the didactical problem must be faced as to how to introduce students to these ideas without trivialising them. Bartolini-Bussi illustrates her approach through a description of a teaching experiment in which historical sources as well as the physical models (most of which had historical relevance) were used. In particular, she notes how the introduction of a conscious *anachronism* fostered recourse to different tools developed in different ages, thus assisting the interrelation of different types of representation.

In "Reconstruction of Meaning as a Didactical Task: The Concept of Function as an Example", Rolf Biehler examines mathematical meaning within the context of the classroom, and its relationship to the landscape meanings of academic mathematical theory. He takes the concept of *function* as an example of complementarity: functions are both mathematical objects and tools for thinking. Beihler uses a detailed meaning landscape of multiple interlinked elements for exponential functions to illustrate how teachers might think about functions. More generally, he shows how a meaning landscape of a concept has a tripartite structure: (a) domains of application, (b) conceptual structure (theory), and (c) tools and representations for working with the concept. Finally, Biehler notes that the systematic reconstruction of the meanings of mathematical concepts remains an important didactical task to be faced.

In "Meaning in Mathematics Education", Ole Skovsmose looks at the similarities between a process of learning and a process of action. He uses disposition, intention, and reflection to define a concept of learning. Thus meaning in learning becomes related to the intentions of the students. When learning is seen as "performed by students", the question becomes "whether or not the students are given the opportunity to act". Does the situation make it possible for the students to perform their learning?

Skovsmose describes the "foreground of the student" as the set of opportunities (amongst all the socially determined opportunities) available that the student interprets as "real" to him or her. In this way, a foreground is a subjectively mediated, socially determined fact and the foreground of the acting person an important source for understanding an action. In a similar way, it is argued that the foreground of a learning person is a crucial parameter in understanding the learning process.

The interlude "Collective Meaning and Common Sense" and the following chapter address common sense in the light of its contrast and interplay with science. Common sense suggests that to know means to justify conclusions that are already formed, whereas in science, to explain means to establish a synthetic relation between premise and conclusion. The polarity of science and common sense in social practice is analogous to the polarity of analytic and synthetic understanding in an individual's cognition. Although science contrasts with common sense, its foundations rest on common sense, as does all human reasoning.

In any human community, some meanings are shared. The common sense of different groups, however, may vary in complexity and in sophistication. Teachers develop an elaborate set of ideas and experiences in constructing their professional reasoning but may nonetheless retain in this professional reasoning some of the characteristics of common sense reasoning. Teacher educators know that in teacher education, scientific knowledge tends to be treated or manipulated within the logic of common sense. In the wider society, expertise is also used by politicians and educational decision makers in a common-sense fashion; that is, by beginning with conclusions and seeking expert scientific evidence to support them. It is therefore argued that not only should the dialectic between common sense and science be acknowledged, but also common sense itself needs to be reshaped and developed.

In "Mathematics Education and Common Sense", Christine Keitel and Jeremy Kilpatrick explore the thesis that sense-making is not only a problem of the individual learner but also a collective process. It has communality, is situated, and is pragmatic. They attempt to broaden the discussion of the relationship between mathematics and common sense by highlighting distinctive aspects of this relationship and by emphasizing the social perspective. Some common-sense conceptions of mathematics educators are challenged, such as the cult of individuality, the universalism implicit in many conceptions of mathematics curricula and assessment designs, and the contradictory assumptions underlying "mathematics for all".

Keitel and Kilpatrick examine some cognitive and epistemological research related to common sense and provide an overview of the impact of social change on the incorporation of common sense into the school mathematics curriculum. They note that common sense has seldom been addressed directly in the literature of mathematics education, and argue that instead of being taken for granted, common sense needs to be on the agenda of mathematics education. Through such work more multi-layered sense-making can be developed, while common sense assumptions can become open for debate.

The interlude "The Construction of Meaning" opens with a vignette describing two students (with a teacher), attempting to solve a geometrical task using dynamic geometry software. The vignette serves to illustrate the diverse factors at play in the construction of meaning within communicative acts: the *agents* (e.g., teacher, students, text, software), each of which has a well-defined, yet changing, status with respect to the object of communication; the *mathematical referent,* that is, the content involved in the communication; the *means and modes of expression* of the content; the *institutional modalities,* such as curricula, schedules, or professional communities that constrain the communication; the *ways of knowing* of the human agents, whose individual histories help explain patterns of interaction; and finally the *ways of interacting* that the human agents have developed in order to make teaching/learning situations more predictable. Each of these factors leads to questions that are addressed in the chapters that follow.

"Making Mathematics and Sharing Mathematics: Two Paths to Co-Constructing Meaning?" by Celia Hoyles includes discussion of the first three factors above in her attempt to synthesize constructionism with socio-cultural theory to form a framework for the discussion of how meaning in mathematics is both constructed and shared. She uses Vygotsky's ideas of tool use to move the focus of attention from the object and how it is constructed to the dialectical relationship of action and thought, and thereby make it possible to merge the idea of software tools with that of thinking tools and communicative tools.

Hoyles argues that through careful design of software and activities, the mediation of student conceptions by software tools can serve to orient students toward a mathematical way of thinking—although not necessarily in the direction foreseen by the teacher. Additionally, Hoyles seeks to show how the computer tool kit can serve as a shared concrete resource, a joint problem space, to manipulate, to mathematize, and to debug. Thus interactions with software exemplify the activity of co-construction, with some parts of the activity left to the computer and others to the students—in fact, what is done by the software and what is left for the students to construct for themselves involve strategic decisions of software design.

The process of webbing and the simultaneous process of abstracting in situ are explored through various case studies of pairs of students undertaking activities with the computer in geometry and in particular engaging with the geometrical construct of reflection. Hoyles argues that the nature of these two foci and how they are operationalized in the school curriculum have an important influence on student interactions, not least because students come to school with a strong notion of reflection, and UK schools approach reflection through activities involving folding and mirrors, with little if any references to properties and axioms.

In "The Hidden Role of Diagrams in Pupils' Construction of Meaning in Geometry", Colette Laborde considers geometry as a theory that, on the one hand, allows us to interpret physical phenomena, while on the other, generates its own problems, questions, and methods. She notes that diagrams play an important role in geometry teaching and that the integration of spatial aspects of diagrams with theoretical aspects of geometry is especially important when one is beginning to learn geometry. Laborde deals with the teaching and learning of geometry in the first part of secondary school, when pupils are faced for the first time with geometry as a coherent field of objects and relations of a theoretical nature. The way diagrams can be used in geometry, the kind of information one can draw from diagrams, and the use that can be made of this information are usually hidden or tacit in teaching. Laborde suggests that diagrams should become a more important component of the learning of geometry, especially when students are involved in problem solving.

Laborde analyzes the relationship between diagrams in a paper-and-pencil or dynamic geometry software environment and the domain of theoretical objects of geometry. She identifies the actions and processes of pupils attempting to construct a solution to a geometry problem, and shows how the existence of geometry software providing dynamic diagrams that are of a different nature than paper diagrams leads to significant changes in the relation between diagrams and theory.

In "What's a Best Fit? Construction of Meaning in a Linear Algebra Session with Maple", Joel Hillel and Tommy Dreyfus argue for a definition of *meaning* within mathematics education that focuses on the constructed shared meaning of a particular piece of mathematics by a group of students. They demonstrate how the conditions of communication influence the construction of shared meaning by examining in detail the transcripts of a linear algebra session of a group of students attempting to solve an assigned problem while working in a Maple lab. The session is replete with different agents of communication including: the students themselves, an observer, the computer and Maple, the classroom teacher (who is not physically present), classroom notes, and the text. Hillel and Dreyfus describe how the role of the different agents changes as the session progresses, paying particular attention to the role of Maple. They also examine the different linguistic and symbolic means used in the communication, and in particular, the role of diagrammatic representations.

The analysis reveals how socially constructed meaning is mediated by different agents whose contributions are sometimes explicit but sometimes not. They also note that these contributions could not have been predicted in advance, thus the process of constructing shared meanings seems to generate its own dynamics.

In "Discoursing Mathematics Away", Anna Sierpinska argues that for a mathematics educator, the mathematical content of communication constitutes the major object of analysis. Unlike a discourse analyst, who looks at general patterns of linguistic communication between people independently of the topics of their conversations or a sociologist who studies social behaviour in various institutional or cultural situations, the mathematics educator investigates communication about *mathematical* topics in *educational* contexts. In particular, the unit of analysis in the study of communication in mathematics education is not a particular utterance or "speech act" but rather whole *episodes of interaction* defined by an identifiable common topic of the exchange—its content.

Sierpinska distinguishes different modes of reasoning in linear algebra: the analytic mode and two synthetic modes, the geometric synthetic and the structural. She then analyses students' difficulties in developing these modes of reasoning by reference to experimental data collected during a long-term observation of five students learning linear algebra. One outcome of the analysis is to note how students can be very creative with respect to modes of reasoning and can, for example, spontaneously use certain intermediary forms that bear characteristics of both the analytic and the synthetic modes of reasoning. These intermediary forms appear as reasonable tools in the acts of explanation or justification, and thus have a certain cognitive value, yet they raise pedagogical problems around the control and validation of solutions, and the development of flexibility in using a variety of modes of reasoning.

Sierpinska also notes that although the type and level of understanding linear algebra that a student develops by interacting with the tutor depend on many factors, the factor of the mathematical contents of the interactions plays a central role. The mathematical content of the interaction, in its turn, depends to a large extent on what the tutor sees as important, and his or her awareness of its possible interpretations. Sierpinska argues that this awareness can positively influence the routines of interaction between the tutor and the student.

The book concludes with the chapter "Meaning and Mathematics" by Michael Otte. His philosophical discussion of mathematical meaning considers the relevance of Pierce's theory of meaning and takes up the discussion of knowledge in terms of "modes of knowing". Thus the active part of coming to know becomes essential. Furthermore, Otte disputes the dualism, so carefully elaborated by Descartes, which has been the basis of many interpretations of meaning: that is, a mental entity, a conception belonging to an internal world, has a meaning if it refers to something from the external world. A conception of knowing becomes more complex when knowledge and meaning cannot be thought of in terms of representations and references, but in terms of activity. A dialectic between the subject (the knower) and the object (what one could know about) is the basis for a "double" constructivism where both subject and object will be constructed and reconstructed in a process of coming to know.

Otte discusses the relationship between the particular and the general, using as an example the relationship between a particular and a general triangle. He claims that mathematical objects do not exist independently of the totality of their possible representations, but at the same time they are not to be confused with any particular representation. Furthermore, Otte emphasises that a mathematical problem is an objective structure that nevertheless has no meaning apart from its possible representations. In this way, he elaborates on the dynamic relationship between the knowing subject and what knowledge can be about, arguing that one pole in this relationship cannot be described independently of the other. Otte brings all these considerations together through an analysis of Pierce's theory of meaning.

In summary, this book sets out to bring together discussions of the multiple perspectives through which mathematical meanings are constructed. The assembled contributions illustrate how meaning has to be sought not only in terms of references to mathematical notions but also in the activity in which students or others doing mathematics are involved. Yet, if we conclude that mathematical knowledge is interwoven with many other forms of knowledge, ideas, and conceptions, then the whole process of constructing knowledge might not only be facilitated but also obstructed by, for example, collective meaning and common sense. Thus we return to the issue of *meanings of meanings*, where conceptual clarification is mixed with empirical observation. We are left with an open question: "What could be the meaning of meaning? " The following chapters may serve to demonstrate why this question is impossible to answer conclusively yet is worthy of investigation.

MEANINGS OF *MEANING OF MATHEMATICS*

Some students find it pointless to do their mathematics homework; some like to do trigonometry, or enjoy discussions about mathematics in their classrooms; some students' families think that mathematics is useless outside school; other students are told that because of their weaknesses in mathematics they cannot join the academic stream. All these raise questions of meaning in mathematics education.

We see, then, that a wide variety of "meanings" can be found in mathematics education: in particular, we note the way in which sometimes "meaning" can be used in a personal sense, whereas on other occasions we are seeking for an agreed, common meaning within a community.

What is the point of doing this? Such a question when posed either implicitly or explicitly is related to other questions such as "Why do you think it is worthwhile?" "How is this linked with your intentions?" In order to discuss the meaning of an action it is important also to consider the notion of intention. This observation indicates that a discussion of meaning within a teaching-learning context should not neglect the complexity of the situation and of the "agents" in this process.

What is the point of making this statement? This question is related to others such as "What do you mean by this?", "What are you referring to?", "What do you mean by this notion?" To discuss the meaning of a conception it is necessary to be aware of the "conceptual landscape" of the agents in the process of communication. Meanings have to be interpreted with reference to the "horizon" of the individual.

In order to develop a theoretical awareness of both the educational meaning of classroom activities and of questions relating to content matter, the notion of "meaning in mathematics education" has to be developed in all its complexity.

1. A VIGNETTE: ON THE DIFFERENT MEANINGS OF DECIMAL NUMBERS

What are decimal numbers? A pure mathematician would be able to give an easy answer: they are elements of a particular algebraic structure. Such a definition, presented fully in formal terms, in no way captures the essence of these numbers for mathematicians, but it is even further from the meaning attached to decimal numbers by, say, shopkeepers or physicists, or by students and their teachers. Most shopkeepers view decimal numbers as a couple of integers used when manipulating money (the currency and its fraction when it exists) whereas for most practical persons they are integers which, together with a comma or a point, are used when giving the measurement of objects (weights or lengths, for example). In the shopkeeper's practice decimal numbers are one way to represent numbers which could otherwise have been represented by integers provided a smaller unit had been chosen. They miss what to mathematicians is a fundamental property of these numbers: their density within the set of real numbers. For physicists or chemists the view is different: the decimal number is related to measurement and the possibility to approximate an "exact value". This use of decimal numbers calls for a theory of approximation, and the way that errors spread over computation. Several other views of decimal numbers related to practice can be mentioned: decimal numbers as seen in schools, decimal numbers as a way of writing fractions, decimal numbers as dynamic representations of real numbers in a computer. These meanings are related to efficient practice, and in this respect they should be considered as contributing to the understanding of the meaning of decimal numbers. When decimal numbers are taught in the classroom, then different practices are developed, raising the question of the meaning of these numbers for the learner, for the teacher, and for the institution, which to some extent fixes decimal numbers as target knowledge. The meaning of decimal numbers for the teacher is an emergent property arising from the interaction between what they mean for him or her as a mathematician, the way he or she views the problem of their learning, and of how he or she implements their teaching. For learners, decimal numbers are tools to solve problems and exercises set by the teacher; and for them the meaning of decimal numbers is shaped not only by this classroom practice but also by the practices they will have met outside the school in their everyday life, as well as the role which decimal numbers play in the school assessments to which they will be subjected.

2. SPHERES OF PRACTICE AND MEANING

In order to facilitate the investigation of the complexity of meaning in mathematics education we shall introduce the notion of "sphere of practice" (SP), which designates the familiar context of functioning of an individual or a given community. This, we hope, may help clarify some of the essential questions related to "meaning" in teaching and learning. One way to characterise an SP is to express the rules, routines, priorities, values, and actions which are attached to it. Alternatively an SP could be defined by the "community" adhering to a common set of rules. That is, an SP is characterised by rules and means that refer to the communication between the "members" of the SP and by rules that refer to the object related activities of the members. So at one level, shopping could exemplify an SP, as would preparing a research paper for publication. Meanings are constructed in SPs, and so could also be characterized by rules, routines, priorities, values, and actions. Meaning relates to all the notions by means of which we try to identify an SP.

In particular, we can observe the difference between the meaning of a concept and of an action. In order to understand the meaning of a concept we could clarify what could be done with the concept. In order to comprehend the meaning of an action we could see what is the intention (implicitly) connected to the action.

Recognising that an individual may be associated with a variety of different SPs allows us better to understand, or even explain, apparent inconsistencies, as well as acknowledging the inherent complexity of "knowledge". To take a simple example, we note that when shopping (a particular form of SP) a person might be involved in quite advanced estimation of proportions, whereas faced with a formal exercise (outside the shop), he or she would be unable to solve a quite similar problem for this is now placed within the context of a different SP. Alternatively, a student may be able to solve classroom problems routinely but not able to use this competence in everyday life contexts. In a way, an SP can become isolated.

SPs can, however, be interrelated in different ways and one SP may come to dominate others. According to the ethnomathematical perspective, many students not only "learn" formal, institutionalised mathematics, but are also told that their already developed "local" mathematics is valueless. SPs might also be integrated. This is what is attempted in many contextualised mathematics examples: familiarity with the context is supposed to relate the mathematics text and so bring together two SPs. (In many cases this hoped-for integration of SPs fails because either the student's understanding of the non-mathematical SP is insufficient, or because knowledge of the latter convinces him or her of the spurious nature of the mathematical task.) Integration of SPs might also be structured by the particular SP established by the didactical contract within the classroom.

Investigating relationships between SPs can, therefore, facilitate research on meaning in mathematics education. Such work can help us better to understand the *mediation of meaning*, filling *meaning gaps*, stimulating the *evolution of meaning*, or the *communication of meaning*, etc.

3. MEANINGS OF (FOR) SCHOOL MATHEMATICS

Society and mathematics educators attribute social meaning to mathematics education. This can be demonstrated in the selection and shaping of the content to be learned; even on the streaming of pupils and the different objectives of education and school types. Great differences can be observed between countries.

Generally it should be our aim to empower students to act in several SPs such as the workplace, everyday thinking, as part of a scientific culture and preparing for becoming part of the academic culture (preparation for college and university education) as well as for democratic life in society. The big differences to be observed between the various SPs in which mathematics occur makes it necessary for mathematics educators to (re)construct the mathematics to be taught taking into account the different needs.

A basic problem lies in the difficulty in communicating, transforming and negotiating the social meanings of school mathematics so that they are shared, accepted or criticised by individual students. This is partly the reason why new meanings for the students are invented, but students may be aware that they are being cheated. Also, it is unfortunate that one obvious social meaning (function) of school mathematics is as a means for the selection and streaming of students and not necessarily as a tool for empowering them, and this fact influences the students' interpretations of the tasks they meet in school and encourages the development of instrumental attitudes and motivation.

4. MEANINGS ASSOCIATED WITH A MATHEMATICAL CONTENT

Mathematical concepts can be defined in a formal symbolic way. Although the meaning of a concept is constrained by its definition, we find that this meaning also depends on the context in which it is used. The role of definitions in academic mathematics and in everyday discourse differs. For the moment we will follow a now classical epistemological position; we take it that the meaning of a mathematical concept lies essentially in the situations it allows us to describe and the problems it allows us to solve in an efficient and reliable way. This viewpoint plays a fundamental role in the design of didactical situations as well as in the analysis of actual practice.

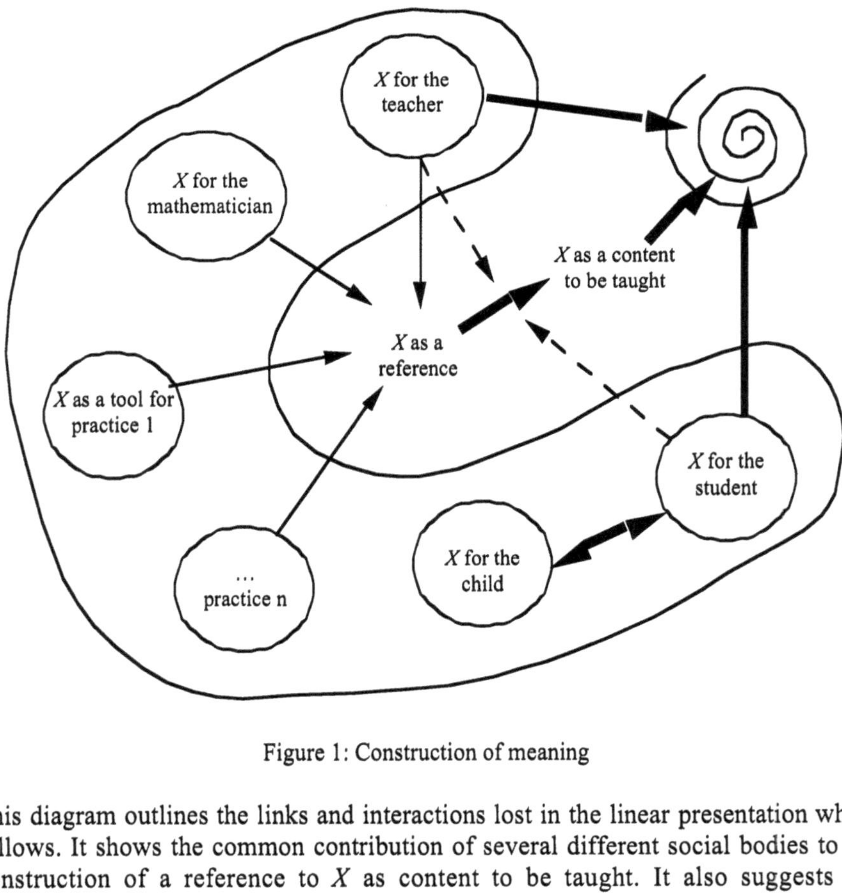

Figure 1: Construction of meaning

This diagram outlines the links and interactions lost in the linear presentation which follows. It shows the common contribution of several different social bodies to the construction of a reference to X as content to be taught. It also suggests the complexity of the didactical transposition which should take into account "teachability" and "learnability", as well as what *X* is from the point of view of the teacher and the learner, and its relation to the child's previous knowledge. In the classroom the teacher is mediating the relation between the student and the content to be taught, through several different processes which could range from guided discovery learning to front of classroom teaching.

4.1 Meaning of X as content to be taught and as a content of reference

In the mathematics classroom, the interactions between the teachers and the students make sense as long as we can identify an "object" of this interaction, which we will call the taught content. By content we mean some piece of knowledge as well as processes, mathematical thinking or other mathematical abilities. In most educational systems this taught content is largely determined by curricula and/or textbooks. A classical position is to see the content to be taught as being the result of a didactical transposition of given mathematical content. But this process of selecting and defining the content to be taught is related to the goals of the participating institutions in society. A broader view, then, is to consider the existence of a *content of reference* as the result of the interactions of several institutions including the mathematicians. The didactical transposition turns this content of reference into content to be taught through a process which transforms the original meaning of either sphere of practice.

4.2 Meaning of X within a sphere of practice

As with "decimal numbers", illustrated in the above vignette, it is important to be aware that all mathematical content holds different meaning in different spheres of practice. A classical view is the following. We have mathematics as a fairly autonomous sphere of practice where mathematicians explore mathematical concepts as objects. In this sphere, the emphasis is placed on the specificity of mathematical problems as the main source of the meaning of mathematical concepts and the role played by the relationships between them. Then we have various other spheres of practice that use mathematical content as a tool for modelling and applied problem solving. Applications outside mathematics add and change the meaning of mathematical content in different ways. It is also the case that mathematics itself is split into different spheres of practice and that there is some exchange of meaning between mathematical spheres of practice and these spheres of practice in which it is possible to recognise mathematical knowledge (even though those involved in such spheres of practice may not be explicitly aware of their using mathematics).

4.3 Meaning of X in the mathematics classroom

The meaning of *X* in the classroom is formed by complex interactions between the teacher and learners with reference to the content to be taught. Furthermore, it is shaped by students' previous knowledge and intentions, the teacher's professional knowledge about learning processes and his or her view of the content to be taught, as well as by such didactical constraints as the time available, assessments likely to shape meaning, and accountability to the content of reference. So, the classroom is the locus of the interaction between several different meanings, among which the gap between the teacher's and the learners' meanings, as well as the diversity of the learners' meaning raises difficult questions still to be investigated.

4.4 Meaning of X from the child's (young person's) and the student's point of view

The student constructs meaning both from the learning process taking place in the mathematics classroom and from knowledge gained outside the classroom: What is usually known as the student's previous knowledge. The student's understanding of the teacher's expectations, combined with his or her assumptions about school intentions as compared to what he or she experiences in everyday life, may well result in the end meaning being at some distance from the one expected by the teacher. Even if students have constructed a certain meaning of a concept, that concept may still not yet be "meaningful" for him or her in the sense of relevance to his/her life in general. Again the meaning of *X* from the child's perspective is not a homogeneous entity but it could depend on the "context" of use (e.g., games with other children, family interaction).

4.5 Meaning of X from the point of view of the teacher

The teacher's point of view of meaning can be explored as the counterpart of that of the learner. It should not be reduced to the meaning of fragments of mathematical knowledge as content to be taught but it should be related to the meaning of *X* as a reference and the meaning of *X* in various spheres of practice.

Moreover, let us consider another aspect. The topic of fractals has recently been introduced in the schools of some countries with a fair degree of success. This innovation cannot readily be explained in terms of the value of this content in professional practices, or in everyday life, or even from the point of view of mathematics itself. What is clear is that we shall not discover meaning for the learning of fractals only by reference to the actual problems in which they occur. Instead, the novelty of the notion in mathematics and the aesthetic interest generated in society by fractal pictures may serve to justify the attempt by mathematics teachers to demonstrate their ability to bridge the gap between school mathematics and the mathematical interests currently to be found in society. This means that meaning from the point of view of mathematics teachers should not be limited to an investigation of their understanding of, and competence in, mathematics, but also of the place and value of mathematics relative to their social status and the way in which they define their professional position.

These views of meaning are actually different meanings of meaning insofar as different methodological tools are needed to explore them, different theoretical frameworks, etc. They insist on several different dimensions of meaning: psychological, social, anthropological, mathematical, epistemological or didactical.

But all these dimensions must not be seen as isolated, one from the other. In fact they constitute a system of meanings whose interactions shape what may be seen as *the* meaning of a mathematical concept.

5. THE SPECIFICITY OF MATHEMATICS

As already indicated, the discussion of the meaning of concepts of and of educational activities is given a special "flavour" when we turn specifically to mathematics.

The invisibility of mathematics in social life seems to be more extreme than in other sciences; and this tends against the wishes of teachers to capitalise on social uses of meaning in order to make mathematics part of meaningful actions. For instance, it is necessary explicitly to make conscious and apparent that certain societal practices and certain technology are essentially based on mathematics. These facts are not obvious: thus, for example, many uses of computers are really based on *mathematical* models but are attributed by the mass media to the "power of computers". Furthermore, mathematics underpins most scientific discoveries and theories. Nevertheless, we can learn and communicate about topics such as astronomy without using mathematical concepts and notation in any explicit and conscious way.

Historical developments have led to pure mathematics taking a form different from that of the other sciences, even physics. A major influence has been the decontextualisation of mathematics or, as some philosophers of mathematics would say, its formal nature. Yet, if mathematics can be seen as the manipulation of "meaningless" symbols according to formal rules, what are the implications for a conceptualisation of meaning in mathematics? There is no doubt that one feature of mathematics has been the development of calculi that do permit the manipulation of "meaningless" symbols.

Ideally there is a shared responsibility in applied mathematics: the validity of a model is not the responsibility of the mathematician but that of the engineer, physicist, banker, etc. How we take this division of responsibility into school contexts is a difficult problem: can we be "epistemologically honest" or do we have to adopt an epistemological standpoint from former times in history? For instance: is geometry the science of space and probability the theory of random situations or do we accept decontextualised versions of the theory (e.g., claiming that probability is the theory of probability measures, and that it is not the responsibility of the mathematician to judge whether or not a certain model is "valid")? To adopt the scientific standard of "rigour" in school mathematics is hardly possible. There is a need, then, specifically to constitute school mathematical standards.

Meaning in mathematics seems to be much more unambiguous and fixed by definitions than in any other sciences. As a result, we may wrongly assume that meaning can always be "given" when introducing a new concept. Yet, of necessity, we have to take into account that meaning has to develop gradually. However, there is a tendency to fix meanings in school mathematics by key applications of concepts—even when no attempt is made to give explicit definitions. How does this relate to the openness of meanings in scientific mathematics, where definitions also have a heuristic role to play?

Developing meaning in mathematics is particularly related to the theoretical nature of mathematical concepts. We cannot ignore that the meaning of a mathematical concept is dependent on a theory whose development will contribute to the meaning. This is also the case in the natural sciences, where the meaning of concepts such as energy and force are theory dependent. Any attempts to teach a physical notion such as energy by reference solely to examples of energy such as electricity, motion, etc. will lead only to the acquisition of an "everyday meaning" of this concept. Only when attempts are made to embed such notions within a theoretical framework will the student enter into the SP of the "professional" mathematician or physicist.

This last example exemplifies the problem of meaning gaps or differences and the problem of relating scientific knowledge, knowledge in school, and knowledge underlying practices outside school. This problem is, however, not specific to mathematics, as the relations and differences between common sense and scientific thought, discussed at length elsewhere in the book, clearly show.

6. QUESTIONS

As a conclusion to this brief analysis, we list some fundamental questions. Some questions are considered in this section or in other chapters of this book; all would reward further consideration.

- How do we deal with the problem of the evolution of meaning both in mathematics, society and an individual?
- How best can we exploit different vehicles for the communication and negotiation of meaning?
- How can "meaning" for teachers and didacticians be developed and links to school mathematics and academic mathematics strengthened?
- How does one justify the place of a topic in the curriculum in a way that it still has "meaning" for students, the academic mathematician, parents and society in general without doing so in a way which ultimately proves destructive of meaning?
- What does "doing mathematics", being a student, a mathematics teacher, a mathematician, mean? How is this meta-knowledge to be developed over the years?
- How can we exploit "common sense", "out-of-school mathematics", etc. without establishing links which will later act as chains?
- How can teachers best explore the meanings which students have constructed?

GEOFFREY HOWSON

"MEANING" AND SCHOOL MATHEMATICS

This book is concerned with the various meanings of *meaning* within the context of mathematics education, and the purpose of this first paper is to take a preliminary look at ideas that will later be developed in more depth. In particular, we shall consider "meaning" in school mathematics and in a wider context, methods whereby attempts are made to give meaning to mathematics, the interrelation of mathematics and its teaching and common sense, and, in the last section, various approaches to the teaching of a particular mathematical topic. The paper was prompted by the study of a sample of textbooks written for Grade 8 students (roughly 13-year-olds) and used in eight European, Asian, and North American countries. This work, undertaken as part of the Third International Mathematics and Science Study, concentrated upon issues relating to the curriculum. The resulting monograph (Howson, 1995) provides inter alia full bibliographical details of the texts studied, but in this paper, where the emphasis is not on comparative studies, references to the exact provenance of examples will not be given.

While reading the texts, my attention was repeatedly drawn to three aspects of *meaning* that school mathematics and its study might have for students, and how these meanings might be affected by the students' use of these materials:

– What meaning is to be attached to the study of school mathematics?

The student has, without any choice, to spend four or five hours a week studying mathematics. What reasons does he or she see for this? To what degree does *school mathematics* possess "meaning" as a coherent body of knowledge and as an activity?

– What do "doing mathematics", "mathematising", and "being a mathematician" mean?

Here we are asking not only about "school mathematics" but also about the way in which that relates to "out-of-school mathematics", including academic mathematics: to a discipline with a history, a present and a future, employed in a variety of forms and ways, and impinging upon all societies and cultures. This, therefore, can be viewed as a form of meta-knowledge, gradually to be developed over the years.

– How can one associate "meaning" with mathematical objects and concepts?

What meaning will, or do we wish, a student to attach to, say, decimal or negative numbers?

These three aspects can be viewed in different ways and, if desired, further sub-classifications can be made. Perhaps most significantly, one must distinguish between two different aspects of *meaning,* namely, those relating to relevance and personal significance (e.g., "What is the point of this for me?") and those referring to the objective sense intended (i.e., signification and referents). These two aspects are distinct and must be treated as such. There is also the need to distinguish between the "meaning" to be attached to mathematical activities (e.g., proving and the organisation of knowledge) and to mathematics as a body of knowledge.

All of these "meanings" will be developed further in this book. Here, we shall concentrate on what might be inferred about them from the study of textbooks, the enormous influence of which on classroom practice shows little sign of diminishing. What is being attempted? What offers the hope of improvement and is worthy of emulation, and what might be done which at the moment is being ignored? This paper will consider such questions—but in only a preliminary manner. It should be unnecessary to stress that studying a text tells us very little about what will happen in an individual classroom and, in particular, that the comparative study of eight texts, one from each country, cannot be extrapolated to provide us with a description of national aims, practices and characteristics. However, textbooks do carry messages (and ones which in many classrooms and countries are very much heeded). For that reason, their study is of considerable value.

1. THE MEANING OF SCHOOL MATHEMATICS

Perhaps the most striking aspect of *school* mathematics textbooks as they have evolved is the way in which they usually omit any description or discussion of "Why?" (If we look at, for example, sixteenth and seventeenth century texts written for the individual student lacking a teacher, then we often find long introductions justifying the study of mathematics and, in particular, the selection of topics to be found in that text.) Moreover, in addition to there being no justification given for the study of the subject, it is now often extremely difficult to perceive any structure underlying the manner in which topics are presented to students: Thus no obvious framework for learning is provided.

Traditionally, mathematics was often taught in compartments that had a habit of being too watertight. Separate textbooks were provided for arithmetic, algebra and geometry, each embodying apparently different approaches and norms. One of the major aims of the 1960s reforms was to move away from this compartmentalised approach and to present mathematics as a unity. In order better to effect this change, certain "unifying concepts" were stressed. In some countries, this meant that much emphasis was placed on algebraic structures. In others, unification was sought through the use of a common language, that of sets, or an emphasis on functions and relations. For various reasons, there is less emphasis these days on algebraic structure, set language, equivalence relations, and so forth. Yet the "unified" presentation of mathematics is now the norm. Only two of the eight countries used separate algebra and geometry texts, but these were provided for high-attainers: Lower-ability students used unified texts. Unfortunately, the acceptance of such all-inclusive texts, written to contain all the mathematics required by a particular type of student for a particular year or stage, has now led to a situation where chapter follows chapter in what too frequently appears a random manner. Students will find it hard to discern any mathematical or pedagogical structure, other than "It's about time for another chapter on geometry". I am reminded of a Schools Inspector who told me that after watching hundreds of lessons he had concluded that most students viewed school mathematics as a huge bowl of stew from which they were regularly ladled a lesson's full.

It is likely that similar things could be said about other school subjects—the common factor being that "meaning" for what the teacher is doing and asking is to be sought not within the subject matter itself but in the socialised institution of the educational system that decrees that the subject is to be taught and examined. Yet those in government who formulate these demands have very little idea what it is desired to achieve by such teaching or what possibilities exist. A certain degree of mathematical knowledge is rightly seen as necessary intellectual equipment for all, but even here motivation will differ considerably. In mathematics the policymakers place emphasis on the acquisition of that knowledge and those skills which will enable students to deal with future utilitarian and vocational demands. However, these same people would place little or no emphasis within school art lessons on developing the skills required for do-it-yourself interior decorating. The obvious utilitarian value of mathematics, extending beyond civic needs to those of the physical, natural and social sciences and to engineering and technology, has succeeded in establishing a firm and unchallenged place for a version of the subject within the school curriculum. As a consequence, the question "Why?"—which must be answered in a much wider context—can be avoided, or answered in national curricula by somewhat platitudinous lists, many items of which show little signs of being seriously addressed. Indeed, I recall an Inspector many years ago specifically saying that students only ask, "Why are we doing this?" when they are bored. The solution to the "Why?" problem, as he saw it, was to keep students usefully occupied and enthralled, rather than to attempt to seek positive justification for classroom practice. Yet, to be fair to him, what he sought was good, enlightened *mathematics* teaching—not a restriction of content to that which could be readily justified on utilitarian grounds. In some ways he took refuge in the myth of the "master teacher" whose knowledge, enthusiasm, and classroom expertise would carry the day without recourse to any direct communication of aims—pupils would recognise that what was being done was all for the best in the best of all possible classrooms. Alas, textbooks are not written with such teacher paragons in mind, and it would seem generally to be acknowledged that some response must be made to the problem of endowing school mathematics with meaning. The universal response appears to be by the provision of motivation through context.

To some extent this would seem a sensible pragmatic solution to what is an extremely difficult problem. Just how easy is it to explain what we are doing and why? At what age can students reasonably be expected to attach "meaning" and "purpose" to the mathematical activities they encounter in the classroom?

Clearly, some solution may be attempted by setting the mathematics to be taught and learned within what to the student are realistic and everyday contexts. Thus, even the single textbook that still bears signs of the 1960s' emphasis on algebraic structure begins every chapter with a contextualised example, and students are told that they will be able to solve or tackle this by the end of the chapter. Another, far less abstract text, has a "green" or "societal" situation, often relating to a "third-world" issue, at the end of every chapter, and data are supplied which give rise to several mathematical questions.

Yet this emphasis on the socially relevant will not by itself solve our problems. To take an example from the former text, determining where a train from *A* to *B* will meet one from *B* to *A* that started one hour later may motivate a student to study the product of rationals (although it is unlikely that adolescents would rank this example very high on a wish-list of problems to which they would like to have solutions), but how is this to motivate, or justify, the introduction within that chapter of such "kernels" as "the rationals form an Abelian group under addition" or that they form a field? Again, a social problem that does not utilise the mathematics developed in the chapter resembles all too well the sugar on the pill. Yet how is motivation to be provided for certain key mathematical concepts? The mathematical notion of proof, for example, is hardly likely to be viewed as one springing from obvious societal needs.

Perhaps it should be pointed out at this point that the eight countries followed very different patterns concerning the differentiation of students by ability and attainment. Some tried to be truly "comprehensive" in that all pupils used the same text (but not necessarily at the same age), and differentiation of aims and content was left to the teacher's discretion. In others, students were clearly differentiated by types of school or clearly defined streams within a school—with different texts for different streams, or were differentiated within a "comprehensive" environment by means of "loose" setting within a particular school (i.e., not according to nationally prescribed norms)—with either different texts or "omnibus" texts to cope with the identified range of needs. Such considerations, which have considerable social implications and are extremely important for textbook writers, are dealt with more fully in Howson (1995). They are very relevant to our considerations, in so far as texts specifically written for average and below average students tend to concentrate almost entirely on utilitarian mathematics and so implicitly provide "meaning" in a more well-defined, yet restricted sense, with consequent social implications. However, there is not space in this paper to develop such matters in detail and, in general, we shall try to avoid becoming too closely involved with the effects of different organisational procedures.

"Meaning" supplied through the motivation of solving "realistic" problems is likely, then, to leave many major areas of mathematics unjustified. Why spend time on learning techniques? Why abstract, prove, or generalise? (This last question is particularly relevant since two countries introduced the notion of formal proof in the Grade 8 texts that I studied. An immediate consequence for the students concerned is a total, and largely unexplained, change in the type of problems and examples that they are set or asked to consider.)

In practice, then, little attempt is made to provide students with answers to such questions. What they are given, almost without exception, is a standard routinised presentation of material in which chapters follow the pattern:

- motivational contextualised example;
- identification of the underlying mathematics which is to form the theme of the chapter;
- practice of techniques in an abstract, decontextualised form;
- application of techniques within a range of contexts.

If, as one suspects, "meaning" for school mathematics is supplied largely by external, instrumental means, then it might be expected that the need for a more clearly defined and explicit description of aims would arise first in a society in which traditional, instrumental pressures and mores are beginning to be challenged or to show signs of breaking down. Whatever the reasons, it is the case that only the USA text makes a specific attempt to set out arguments for studying school mathematics. The other texts take the matter to be self-evident, as something to be left to the individual teacher, or as implicit within the choice of topics and examples. Let us look at the arguments to be found in the USA text.

First, we notice that the explanation/discussion of "why?" is separated out from the main text: It comes in the preliminary pages. "Mathematics is valuable and interesting", we are told, and students are further informed how the book will help them to explore and discover more of its "wonders". First, the book will help build "math power" by developing problem solving and critical thinking—solving problems will do more than just supply answers, "you will learn how to think mathematically". The student will also learn to build connections between mathematics and "the real world, between math and other school subjects, and between different topics within mathematics". This math power will be developed by working on number sense and using data, and by using calculators and computers. The ability to do one's best will be aided through having a positive attitude, building understanding, and learning ways to study independently. Active learning will be experienced through activities with materials, reading, writing, working in groups, exploring (looking for patterns, checking out hunches and trying different approaches to problems). All this will result in the reader "enjoying mathematics".

This is only a brief summary, but in practice little more is actually said. It would, then, be only too easy to make fun of such a list. Yet, this is the only text in the sample that attempts to provide an explicit rationale for studying mathematics. We note odd omissions (e.g., any specific mention of geometry), but it is perhaps of more significance to ask to what extent are the goals expressed here actually sought?

Before considering this further, two points should be made. First, this is what I term an omnibus text: It contains far too much material to be covered by one student in any one year. The teacher can select from it a suitable course for a high attainer and also one for the less able student. There is no guarantee that the latter would be given very much opportunity to develop, say, problem-solving skills. The second is that this is not a "bad" or "hopeless" text (on a comparative basis). It can be criticised on a number of grounds, but in many ways it probably "tries harder" than any other of the texts to enlarge the types of learning opportunities offered to students.

Yet what one finds in the main text is a (fairly) rich collection of activities (exercises, problems, and explorations—plus plenty of straight-up-and-down traditional exercises that could provide a year's course for the students of a not-too-innovative teacher). There is no discussion of the part that these activities or the mathematics upon which they are based will play in the student's development, or of any pedagogical or mathematical reasons that led to their inclusion. If the student does hone his or her problem-solving skills, it will only be through introspection or the direct intervention of the teacher. Certainly the book does not attempt explicitly to help the student build up a framework within which he or she can reflect and organise experiences and knowledge purposefully. To this extent it is, of course, far from unique—indeed, none of the books I studied attempted to meet this aim. Yet this would appear to be possibly the major challenge that we face.

We all want students to reflect on their learning, but how are we to help them form a framework within which this can usefully be done? Clearly, students will not necessarily reflect in the same way and will assign their own meanings to mathematical ideas and concepts. This is an enormous problem to tackle, yet, unfortunately, I can discern no attempt whatsoever in any of the books to deal with it, other than by the insertion of checks to see whether or not the student has learned a particular fact or procedure. If students do reflect, it may well be on "what new facts have I learned?" or "what new techniques or skills do I now possess?" There is a tendency, then, for books to fragment mathematics rather than to reveal relationships between constituent parts. (The mere acquisition of skills and knowledge can, of course, provide "meaning" and motivation for those students who are "successful". What we must ensure here is that such "satisfaction" arises from activities that have a sound mathematical and intellectual basis.)

To a large extent many of the points on which I have touched arise because of the somewhat ambivalent relationship which exists between what we might term "school mathematics" and "other mathematics", that is, mathematics as viewed by those academics who term themselves mathematicians, and as it is employed in commerce, industry, and other disciplines. In the 1960s, there was a strong move to relate school mathematics more closely with academic mathematics. It could be argued that since then the pendulum has swung too far towards identifying school mathematics with contemporary usage of mathematics in day-to-day life and as a service discipline. This particularly affects the manner in which topics are approached and learned. What might form an appropriate balance is worthy of serious and specific consideration. However, it would seem essential that any student who specialises in mathematics at school should be aware of how an academic mathematician approaches the subject, if only better to inform decisions relating to further study, and I should argue that so far as possible, other students should be given some introduction to the "culture" of mathematics as part of their general education.

This last point leads naturally into the next section, although it must be pointed out that what is written in that and other sections will still have relevance for this first one.

2. THE MEANING OF MATHEMATICS

What impression of mathematics will the student gather from reading the school texts that I studied? What attempts are made to enlarge the students' perspectives on mathematics?

The overriding impression is that textbook authors have attempted to answer (within the framework of national curricula composed with a similar orientation) the following question:

– What mathematics should students know about?

The non-commutability of words within an English sentence is clearly demonstrated by the fact that very little attention is paid to the related and extremely important question:

– What should students know about mathematics?

Of course, *mathematics* has very different meanings in the two sentences. The first presupposes that mathematics is viewed as a collection of results, techniques and, nowadays to an increasing degree, processes. The second sees mathematics not only as a "collection", but as a growing organism—some kind of multi-rooted banyan tree, still growing in all directions with the help of a great variety of arboriculturists working, albeit, with very different motives.

It is sad, but students in most countries would not learn that they are studying an expanding subject, having a rich history that can be traced in many different cultures, and that has been developed by men and women in response to a number of challenges, roughly categorised as societal or intellectual. References to history and to historical figures are few and far between, and are slanted too much by nationalistic considerations. For example, in the books I studied, I can recall no photographic or other representation of a mathematician later than Descartes (1596–1650), and no mention of one later than Abel (1802–1829). (No prizes will be awarded for guessing the countries of origin of these two texts!) I can recall no mention of a woman mathematician. (Yet, strangely, one text had notes on some twentieth century scientists, two of whom were women. What message is being passed on by that example?) One Japanese text had a separate "topic corner" devoted to the history of arithmetic in China, and one of the European countries built in occasional references to the contributions to mathematics of the Egyptians, Hindus and Arabs.

Yet even such references, welcome as they are, tell us nothing of mathematics as it has grown in the 20th century; not only of the mathematicians who have contributed to its advances but also of the way in which these have influenced our lives. Mathematics as exemplified in texts, even the better ones, is almost entirely oriented towards the solution of problems and exercises. Even those "topic corners" that do exist—i.e., pages separate from the main text of the book that introduce problems or topics that might lie outside the narrow confines of the prescribed syllabus—are built around mathematics that can be both readily comprehended and which can be converted into appropriate tasks. There is nothing to read, or to appreciate which does not lead to a problem or exercise to be solved. The idea that it might be educationally advantageous to describe the results of some mathematics, without the students being able to comprehend all the content needed to establish a proof, has either not been considered or has been rejected. Clearly, arguments could be put forward for rejection—most importantly, it would take time away from the practice of those tasks on which, eventually, the students' "mathematical attainment" (and, even, teachers' professional competences) will be assessed. It would also not be easy to find convincing examples and explanations, and their inclusion in texts would of necessity send up prices and thus lead to a probable loss in sales and revenue. Yet our continued neglect of such problems, *at all educational levels,* has meant that the vast majority of teachers have very limited views on what mathematics might "mean" and that textbooks transmit a similarly restricted message.

Of course, some of the concepts we are trying to illustrate are extremely complicated and we shall not always be successful in our efforts to communicate them. Many of the references to history and culture will fall on deaf ears. Yet even if this should sometimes be the case, what of value is lost, compared with the benefits that such an approach might well bring for many students—and for the future of our discipline? For history must not only be seen as an aid to motivation: a way of making mathematics more palatable for those students having a historical, social or philosophical bent. It allows the social dimensions of mathematics to be appreciated and, in so doing, adds to the students' general awareness.

Certainly, I feel that a historical, multicultural, and humanistic approach is more likely to be successful if it permeates a book rather than if it appears in chapter-size chunks at regular intervals—as was done, for example, in a serious and professionally able way in the 1960s East German (DDR) texts. There are plenty of pegs on which we can hang cultural and even contemporary references. Thus, to take an example arising in one of the texts I studied: it was stated that any odd number greater than seven could be expressed as the sum of three odd primes, and students were then asked to show that this was true for the odd numbers between 81 and 91. The exercise has several appealing pedagogical features. What it does not reveal is that the result is far from a God-given fact. Actually, we still have no proof that the result holds for all odd numbers. What might astonish some is that we are certain only that it is true for very large numbers (established as recently as 1937) and for those small ones we can check by hand or computer. Bridging the gap between "very large" numbers (which "meant" having at least six million digits in 1956; tens of thousands of digits by 1989) and the small ones still continues. Here we have an opportunity to demonstrate not *how* something is proved mathematically, but what a proof means to a mathematician, how mathematicians continue to search for them, and that *there still are active mathematicians.* (Moreover, that the early results emanated from Russia and the 1989 one from China demonstrates the international nature of mathematics.)

If mathematics has a history, so does mathematics education. It was interesting, then, to see that two books made use of examples drawn from old school texts. It would be pleasing to say that this serves to illustrate advances made in the teaching of our subject. To some extent this is true, but I suspect that many differences arise more from the packaging, the use of colour, typography, and so forth than from marked pedagogical gains. Yet changes of emphasis can be observed as well as, in the case of contextualised examples, many interesting social and historical changes. Again, though, opportunities to draw attention to the fact that mathematics education itself is not fixed in time but is influenced by social, intellectual, and political demands were not taken. For example, neither book mentioned what types of students (drawn from which social classes and by what means) would have been using the historical texts from which the pages were reproduced. Again, the example offered the chance to educate, rather than merely presenting a new context within which mathematical tasks could be set. There was no need for long explanations; it would have sufficed to supply data for the observant student or which might be used by the teacher to foster a questioning attitude.

3. GIVING MEANING TO MATHEMATICS

Mathematics educators often see it as their primary task to help students create "meaning" for the concepts they handle. This is frequently done through the use of metaphor, e.g., equations are associated with balances, or through contextualisation, e.g., negative numbers seem almost invariably to be introduced *via* thermometers and temperature. As Thom (1973) expressed it, "meaning in mathematics is the fruit of constructive activity" and depending upon the activities to which students are exposed, or in which they engage, different meanings will be ascribed to mathematical objects. There is no obviously shared semantic universe. Indeed, my wife tells me that what was for her perhaps the most memorable school mathematics lesson consisted of the teacher asking members of the class what a "point" meant to them, and the subsequent discussion and argument.

Yet, serious questions can be asked about the means we use in our attempts to provide "meaning" for mathematics. For how long does a particular metaphor help? To what extent can false, limited, or overcomplicated representations hinder? We shall return to such concerns later; here it should suffice to point out that the "meaning" we assign to a concept could be considered to be that equivalence class of representations which we have constructed around it. Thus, for example, around the number "5" we should wish to have both ordinal and cardinal representations; similarly, we should want students to be introduced to a variety of representations of a fraction, both as a rational number and an operator. Without such a planned variety of contexts, then any "meaning" which students construct is likely to be deficient.

I shall not dwell on such arguments here: They are covered more extensively elsewhere in this volume. However, certain points would seem to be worth making at a simplistic level:

- If mathematicians do not share a semantic universe, then their wish to attain rigour can only be granted through the elimination of meaning—the creation of purely formal systems. Their wish to obtain a communality of "meaning" is sought by building definitions upon commonly accepted problems and fields of application.
- The degree to which "text" can actually convey "meaning" has been a source of debate for many years (see, e.g., Otte, 1986, in the very first BACOMET volume). The remarks made above would suggest, then, that if we are to look for meaning in a school text, we should pay particular attention to two aspects:
- Can the informed reader (the teacher, the mathematics educator) identify clear "meaning" which might be spotted and absorbed by students?
- Is the range of representations and problem situations sufficiently rich to enable students to ascribe "meaning" in a "meaningful" way?

(Somewhat paradoxically, I find that these conditions are rarely met in those attempts to replace "traditional" texts, by what I should term "activity-based" books.)

Learning through metaphor, immensely valuable though that might be, tells us what something resembles, not what something is—and the temptation to injudicious extrapolation from the metaphor might prove too great. In the section "Directed Numbers" we shall look in some detail at how the texts studied deal with negative numbers and of their use of metaphors; and in this connection it is interesting to recall how, in the late 1700s and early 1800s, many in England still looked upon such numbers with considerable misgiving. Indeed, in 1796, Frend, a Cambridge mathematician, produced an algebra text, *Principles of Algebra,* in which he avoided their use. He argued that "multiplying a negative number into a negative number and thus producing a positive number" finds most supporters "amongst those who love to take things upon trust and hate the labour of serious thought, [for] when a person cannot explain the principles of science without reference to metaphor, the probability is that he has never thought accurately upon the subject". Frend's son-in-law, the better-known mathematician De Morgan, was to write in his *On the Study and Difficulties of Mathematics* that "the imaginary expression $\sqrt{(-a)}$ and the negative expression $-b$ [...] are equally imaginary *as far as real meaning is concerned.* [One] is as inconceivable as [the other]".

Learning through metaphor and contextualisation can prove restricting, unless one is prepared eventually to move towards a more abstract view. Yet if mathematicians had nothing to fall back on but the strong syntax of a formal system, then the subject would not have developed to the extent it has. Whitehead and Russell's *Principia* (which is mathematics bereft of meaning) is not fertile soil.

Algebra presents us with especial problems. Geometry is strong in meaning. (The class could discuss the "meaning" of *point,* but that of "*x*" might have created more difficulties in getting started.) Thom summed this up by remarking how geometry was rich in meaning but weak in syntax (although it allowed a psychological widening of the syntax while still retaining the meaning always given by spatial intuition). In contrast, algebra was weak in meaning but rich in syntax—"the meaning of an algebraic symbol is established with difficulty or is non-existent". Indeed, the power of algebra lies in its abstractness and the way in which it can be handled autonomously independent of meaning, thought, or content (as pointed out by Whitehead, Freudenthal, and others). That students can obtain solutions to problems, particularly geometric ones, through the use of algebra and coordinate geometry has given rise to pedagogical arguments for over two centuries. In the early 1800s, there were strong objections in many European countries to the teaching of coordinate geometry on the grounds that it detached meaning from mathematics. Simson, the geometer and editor of *Euclid,* writing in the mid-1700s, thought "[algebraic analysis little better than a] mechanical knack [in which the student proceeds] without ideas of any kind to obtain a result without meaning". In some ways this provides a pre-echo of today's controversies concerning the use of calculators. What was considered "meaningful—such as long division, with its apparent emphasis on an understanding of the place-value system and which also served a useful purpose as a preparation for work with polynomials—has been replaced by a mechanistic method which supplies answers without any seeming intellectual effort on the student's part. Are we, then, interested only in the mathematical "end product" or in the pedagogical journey there and the insights gained en route?

The mathematician employs two principal means to generate meaning, both of which are used to some extent at school level. The first is that of forming geometrical models and so utilising spatial awareness and intuition. Geometrical interpretations take on a number of forms. A key example from history is that of complex numbers. Various attempts were made to interpret these in a geometrical sense before Gauss in 1831 achieved this in a way that successfully modelled both elements and operations and finally gave meaning to what Napier had previously described as a "ghost of a quantity". Gauss, himself, was to claim that as a result of the geometrical representation the "intuitive meaning of complex numbers [is] completely established, and more is not needed to admit these quantities into the domain of arithmetic". Again, geometrical models for hyperbolic plane geometry, such as those by Beltrami, Klein, and Poincaré, have the psychological and logical effect of binding together the new and "abstract" with the old and seemingly concrete Euclidean plane.

This "binding" is also to be found, indeed is the goal, of the second approach: that of describing or defining the "new" in terms of the accepted. Thus by defining the integers in terms of equivalence classes of ordered pairs of naturals, we overcome the objections of Frend and De Morgan. Similar devices allow us to construct the reals, and again, by means of ordered pairs of reals and suitable definitions of the operations of addition and multiplication, we arrive at the complex numbers.

Yet, although such methods give confidence to mathematicians by adding a degree of legitimation to their work, their pedagogical use is limited. In the 1960s, when attempts were made to introduce axiomatics into school mathematics, frequent use was made of what were referred to as "parachute postulates". These were attacks on the problem from the rear, in that they were not framed because they were natural assumptions (in the manner of Euclid), but in hindsight because they allowed one to get around what in the normal way of things would be points of difficulty. In a similar way, the definitions used in building the models mentioned above are very much "parachute" definitions. They only make sense if one already appreciates what they are intended to legitimate. To define the product of (a, b) and (c, d) as $(ac + bd, ad + bc)$ in one case, and in another as $(ac - bd, ad + bc)$ appears completely arbitrary unless one is already familiar with operations in the system to be modelled. Such procedures are mathematico-logical rather than pedagogical. It was, then, surprising to find a garbled form of this approach still used in one text. In this, rationals were defined as equivalence classes of fractions—however, without clear definitions of the terms employed. The pedagogical and mathematical aims of such an approach were not apparent.

Geometrical interpretations are far more widely used, and we shall have cause to draw attention to some of these in a later section.

4. MATHEMATICS AND COMMON SENSE

> *Common sense*: average understanding; good sense or practical sagacity; the opinion of a community; the universally admitted impressions of mankind. (Chambers' *Twentieth Century Dictionary*)

In his "China Lectures", Freudenthal (1991) asks whether or not common sense is not the primordial certainty, the most abundant and reliable source of certainty within mathematics. He goes on to point out the extent to which number and elementary geometry (e.g., ideas of similiarity) are grounded in common sense. Children acquire number, and assign meaning, in day-to-day activities—various representations are shared by the community as a whole and, as a result, ideas related to number qualify as "common sense".

That this does not solve all the problems of teaching arithmetic is readily accepted—misconceptions still abound. Nevertheless, the approaches to be observed in the textbooks studied would usually appear to be based on the principle that all we are doing is trying to codify common sense and to extend commonly accepted notions.

Of course, *common sense* is far from being a scientific term: It is a vague, culturally dependent, but nonetheless extremely valuable concept. In essence, it provides us with a means to talk about mathematics, and with a rudimentary, one-way form of logical reasoning. It is distinguished by the way in which it depends upon evidence, accepted truths and conventions, and upon "innate" operating systems of perception, meaning and understanding. There is no doubt that it provides a powerful tool for survival in social life. It cannot be denied, then, that common sense is something that educators must try to develop in students and, conversely, something on which we must draw in our teaching.

Yet there comes a time when the link between common sense and mathematics breaks down—and this gives rise to tensions in the books studied. Frend's objections to negative numbers were based on common sense: How could multiplication of two of these mysterious objects yield something with which he was familiar—something which was part of common sense? When Gauss suppressed his findings on hyperbolic geometry because, as he wrote, he feared the shrieks that they would elicit from the Boeotians (a dull-witted Greek tribe), he surely meant that many of his fellow mathematicians would reject, as an affront to their common sense, the assumption that given a point and a line in the plane then there could possibly be more than one parallel to the line through that point. Poincaré found Cantor's set theory "a disease" from which he hoped mathematics would recover: The reason, in Du Bois-Reymond's (1882) words, was that "it appears repugnant to common sense". Clearly, the theory of infinite sets *does* contradict common sense. Let us take a simple example. Suppose there are football players on a field, some in red shirts and some in indigo. If I ask them to form a line according to the rule that between every two in red there must be one in indigo, and between every two in indigo one in red, and then note that the line begins and ends with a person in red, common sense tells me that there is one more person in red than in indigo. But what happens on the number line in the closed interval 0 to 1? The interval begins and ends with a rational. Between every two rationals there is an irrational and a rational. But there are not more rationals than irrationals: Indeed, the number of rationals is insignificant compared with that of the irrationals. What has happened to common sense?

Mathematics should not be confused with, or, indeed, constrained by common sense. The latter, if it is to become genuine mathematics, must be systematised and organised. This seemingly obvious remark can create problems for the textbook author or curriculum developer. Let us take one simple example based on exercises to be found in one of the series I studied. Students were provided with photographs and asked, on the basis of the plans provided, to identify the church photographed, or, using a town plan, to say from which point a particular photograph was taken. These I see as useful activities to develop "spatial awareness" (whatever that means!), which mathematicians will wish to draw upon. Yet in these sections the textbook had no "kernels" to offer, that is, no mathematical definitions, results, or procedures that students might identify, learn, or follow. Rather, opportunities were provided for students to use their "common sense" in order to extend this still further. It was hoped that in so doing they would develop desirable, idiosyncratic traits, the logicality and systematics of which were, however, never discussed or assessed. Is this the best that we as mathematics educators can offer? Strangely enough, 1960s textbooks carried within them a language of systematisation that might be applied to this problem. Looking at a town plan, we could consider the set of all points from which, say, the spire of St. Peter's church would appear to be on the left of that of St. John's. Consideration of the intersection of various such sets would lead to the desired answer. In the 1960s, such context-based questions were rarely, if ever, set. Now perhaps there is a danger that context and a dependence upon common sense are driving out mathematics. There is a nice balance to be observed.

Already, then, we begin to see certain problems arising in connection with "mathematics and common sense":

a) Although founded upon common sense in its meaning of "universally admitted impressions of mankind", mathematics is more than common sense. Indeed, the latter can be a constraining force on the learning, comprehension, and development of mathematics.

b) In our mathematics teaching there will frequently be a need for us to develop our students' "common sense" (i.e., bring a student up to "average understanding" as in the example just cited on spatial awareness) in order for him or her to make progress in mathematics. This may be a legitimate part of mathematics education, although perhaps scarcely qualifying as "mathematics". Moreover, children from different social communities will bring with them different types of "common sense", for this is not a universal constant; indeed, it has been described elsewhere as "local knowledge". Discussions of "ethnomathematics" might well gain, then, by focussing more on the issue of the common sense peculiar to a particular society.

c) Mathematicians as "a community" have their own brand of common sense. A major aim of mathematics education is to develop this type of common sense in students—to add to what they consider to be normal mathematical behaviour, to develop that knowledge and those methods of thinking which are often ascribed to "(mathematical) common sense". (Does, for example, "It is obvious that..." really mean: "Given my knowledge of mathematics, it is (mathematical) common sense that..."?)

d) If the Greeks (in the case of irrational numbers), mathematicians and physicists such as Frend and Lazare Carnot (negative numbers), Poincaré and Du Bois-Reymond (set theory) had difficulty in accepting new ideas that contradicted their common sense views, then it should not surprise us if our students have problems.

The difficulties mentioned in (c) and (d) were seen to arise quite clearly in the case of a student observed by Sierpinska (2000) and in video by the BACOMET group. He, when asked to "prove" that in an abstract vector space there can only be one zero vector, was totally puzzled. To him, based on the study of a particular geometrical representation of a vector space, this was clearly a matter of common sense. What, he was essentially asking, are we trying to prove, and what need is there for proof? (Is the concept of proof one generated by common sense, or is it essentially lodged firmly within mathematics? Again, a problem raised—but in my view inadequately answered—in the two series that introduced the notions of formal proof in Grade 8.)

What was the cause of the student's confusion? To what extent was it due to a previous "reliance" and emphasis on common sense and on experience based on restricted contextualised representations of vector spaces on which "common sense" thrives?

There comes a time in mathematics when the constraints of common sense must yield to the demands of structure—whether this accompanies the introduction of new types of number and the operations upon them, or of algebraic and topological structures *per se*. In some ways this parallels developments in the foundations of mathematics, from Frege, who attempted to build mathematics upon the foundations of a "common sense" logic, to those of Hilbert and the development of formal systems. Thus, for example, in 1919, Hilbert stressed that the concepts of mathematics were built up "systematically for reasons that are both internal and external" (quoted in Rowe, 1994). Common sense supplies "external" motivation, but we must look to mathematics for the internal reasons.

Making the change from a view based on common sense to one that takes into account the internal structure and demands of mathematics will never be easy, and attempts to disguise what is happening will almost invariably lead to confusion in the students' minds. In the next section, we shall see how this problem arises and is dealt with in the various texts in connection with the multiplication of integers.

Other problems arise when we do not distinguish clearly enough between the world—and our commonsense conceptions within that—and the abstract model of it that we form within mathematics. In his *Pathway to Knowledge*, Recorde (1551) tackled this problem head on:

> A *point* [the spelling has been modernised throughout this extract] or a *prick* is named of geometricians that small and insensible shape which hath in it no parts, that is to say, neither length, breadth nor depth. But as the exactness of this definition is [more suited] for only theoretic speculation, then for practice and outward work (considering that my intent is to apply all these principles to work), I think it [more suitable] to call a point or prick that small print of pen, pencil, or other instrument which is not moved nor drawn from the first touch.

Similarly when defining a line, Recorde was led to observe how geometers "in their theories (which are only mind works) do precisely understand these definitions". Whether he shared our appreciation of the distinction that he drew between the abstract nature of the geometer's system and the artisan's real world of which it is a model is, of course, doubtful. Nevertheless, it demonstrates the care with which he approached his task. Perhaps at the other extreme to Recorde, insofar as it confuses abstract mathematics with the real world, is an example I saw recently which sought to bring sense to irrational numbers through contextualisation. It was a proposed test item which described a clock having a minute hand of length $\sqrt{5}$ cm and an hour hand of length 2 cm. The students were asked to state whether the distances between the tips of the two hands at 12, 6 and 9 o'clock were rational or irrational. The question demonstrates a degree of bizarre ingenuity. The outcome, however, is more irrational than the numbers involved, for irrational numbers are part of the mathematician's abstract world: they are not measuring numbers in the real world. Indeed, it is difficult to imagine what an irrational measurement could mean in physical terms. (Incidentally, what clock has minute and hour hands moving in the same plane?) Here, then, we have a misguided attempt to supply meaning—driven not by an appreciation of the demands of mathematics, but by a belief that meaning and motivation must be provided through "real-world (!)" contextualisations.

Recorde's concerns over the meaning of *point* give rise to other considerations of meaning and common sense, or, perhaps better, common usage. Many mathematical words that we use are also to be found in everyday language—sometimes with slightly different meanings. There are many examples of *faux amis*. This is not merely a question of translation between different national languages (although I still treasure memories of an international meeting at which *corps infinis* (infinite fields—a term that in itself carries no meaning in either English or French—was translated as "unfinished bodies"). It concerns translation between relatively imprecise day-to-day language and that of mathematics. For example, the word *similar* gives rise to many problems. This is particularly the case because the everyday meaning subsumes the mathematical one, and so the learner is not faced with a conflict of "meaning" that might force a clarification of issues. The meaning that the child constructs for him- or herself is all too likely to be the wrong one. Many such examples could be given. Others could demonstrate how quantifiers, which play such an important role in the writing of mathematics with precision, are used in a sloppy, and often essentially meaningless fashion, in common speech, textbooks, and even official documents.

Yet to what extent are the mathematical terms we use specifically chosen so as to convey meaning to students (as well as interpreters), that is, words rooted in common sense or common usage and that help bring meaning to the concept with which they are to be associated?

Here, again, we have an early example of the recognition of this problem in Recorde's books. In these, the first major mathematical texts to be written in English, he sought to provide meaning by discarding classical terms based on Latin or Greek and, instead, used such terms as *touch line* for tangent, and *threelike* and *twolike* for equilateral and isosceles. He was soon to abandon this attempt to overcome scholarly usage for "if [I were to do so] many would make a quarrel against me, for obscuring the old Art with new names". The result is that English and North American students have still to battle with names which, in the absence of any knowledge of the classical languages, carry little implicit meaning. In some Northern European countries, those who sought to introduce new vernacular terms were more successful. Thus, it was interesting to note that in the Norwegian texts there are many terms such as *likebeint* (equal limbs) for isosceles and *likesidet* for equilateral. Here, then, mathematical terms carried meaning based on common sense or common usage. At the other extreme, an example of the no-meaning approach was provided by the USA and English texts. For in British English and in North American English, the two terms *trapezium* and *trapezoid* have reversed meanings. (In England, a trapezium is a quadrilateral with one pair of sides parallel, and a trapezoid one with no sides parallel. In the United States, the meanings of the two words are reversed.) Yet etymologically the "best" candidate for a "trapezoid" (table-shaped object) would probably be a rectangle!

These examples provide us with yet another indication of the interaction between mathematics and common sense. Many more such can be found in the literature.

5. DIRECTED NUMBERS

In this final section, we shall consider the manner in which directed numbers and the operations on them were introduced in the texts studied. Although, the books may well have been superseded in the intervening years, they still give a fair indication of how operations on directed numbers are taught in many countries. This topic was one common to most of the texts although, in some cases, students had met the addition and subtraction of integers in earlier years. A later study (Howson, Harries & Sutherland, 1999) of primary school texts from 10 countries also indicated that in half of these ordering a set of negative numbers (using a number line or scale) was taught by Grade 5. The introduction of the multiplication of directed numbers is important in providing us with a situation in which common sense fails and the opportunity arises to introduce the notion of a definition constrained by the internal structure of mathematics.

When considering these brief descriptions, readers might find it helpful to relate them to Vergnaud's (1983) notion of a "conceptual field". Briefly, such a field comprises a set of situations whose mastery demands a variety of concepts, procedures, and symbolic representations closely connected with each other. How rich a "field" will students have?

First, we must note that in one country, pupils in the lower half of the ability range did not study operations on integers in this grade. Vector shifts were first studied on the number line for positive integers. After this, negative integers were introduced via temperatures and the thermometer. The number line for integers was then developed and attention paid to ordering. Questions were set on temperature differences, and these were repeated in an abstract form using the number line. Some elementary work on coordinates was attempted (x-coordinate positive, y positive or negative). The corresponding text for higher-ability students asked for little more in the form of content but set rather more difficult questions (and, in general, demanded much more in the way of algebraic technique). Another country separated pupils according to ability and offered three graded series of texts. The text for the median student did not deal with the multiplication and division of integers. Indeed, even subtraction was optional for, according to the teacher's guide, this "is only really needed by those who go on to a certain level of algebra". (This remark gives rise to many serious questions for curriculum developers, e.g., When is it reasonable to decide that students will not go on "to a certain level of algebra"? What should govern that decision? To what extent should students be taught material which is extremely unlikely to be developed mathematically, or used in "real life"—might other mathematical topics prove of more value to them?) Negative numbers were introduced through temperatures and addition and subtraction via shifts in temperature—a feature of the vast majority of the texts studied. Interestingly the text used an "upper" minus sign ($^{-}1$) to denote a negative number, thus distinguishing between the symbol for the operation of subtraction ($-$ 1) and that for a negative number. In the series for high-ability students, addition and subtraction were developed in a similar way, but kernels were expressed in a general form using the variable n rather than by listing specific examples. The exercises were also more formal than in the book for the median student. Multiplication of integers was introduced by considering the way in which the scale factors of enlargements (dilatations) combine. This, a method unused elsewhere, is straightforward when the domain is limited to the positive numbers and removing that limitation, and assuming commutativity, allows the product of a negative number by a positive one to be defined in an obvious way. Similar "extensions" of meaning led to the presentation of rules for the signs of products. Exercises were given on substitution into formulas, calculator use, the solution of equations, and equations and graphs.

The last set of exercises was contextualised; the preceding ones mainly abstract. I did not spot any contextualised exercises on multiplication or division.

One other country's approach was unique in that it still showed the influences of the 1960s reforms (an influence that has since disappeared). The book began with a chapter "Operations on the Integers". After a preliminary, contextualised question referring to numerical codes, it switched to an abstract, formal approach. The set Z was defined, and the number line introduced. Next followed a definition of the absolute value of an integer, together with five worked examples and four "activities". Kernels then followed thick and fast without any attempt to justify them: "The sum of two integers of the same sign is another integer with that sign and with absolute value equal to the sum of the absolute values of the summands"; "the sum of two integers having different signs is another integer whose absolute value is the difference in the absolute values of the summands and which has the same sign as the summand having the greater absolute value". Commutativity, associativity, and the existence of a neutral element and of inverse elements under addition were then commented upon, and by the fourth page of text we had arrived at the statement that Z is an additive Abelian group. The eighth page of text told us that Z is a ring. There were no contextualised examples and comparatively few formal exercises. Definitions and other examples of the algebraic structures were not given.

Another country made great use of the calculator as a *deus ex machina*. The addition and subtraction of negative numbers had been introduced in an earlier grade. Now the product of a positive integer and a negative number (note that the negative numbers were not restricted to integers) was obtained by repeated addition and using a calculator, and the rule for the sign of the product of two numbers inferred. The product of two negative numbers was obtained using a calculator, and an appropriate rule suggested. This approach was then repeated for quotients. There were a handful of formal exercises to be done without the aid of the calculator and a multi-faceted exercise on graph plotting which involved not only the use of negative numbers but also their multiplication. After a formal review of kernels, there was a collection of exercises that included a number of "well-concealed" repetitive exercises involving magic squares and so on, and a number of items where students were given four numbers (positive or negative) and challenged to combine these using the four arithmetical operations to obtain a fifth given number. Some algebraic examples were included at the end of the chapter but only the higher-ability students would presumably have reached them.

In another country, the integers were introduced in Grade 7 by "conventional" means—that is, temperatures, heights above and below sea level, and so forth—leading to the number line for the integers. Addition and subtraction were then introduced via vector "shifts" on the number line and multiplication by taking scalar multiples of such vectors. Exercises were many and abstract, with emphasis being placed on the rules for dealing with the "product of signs". In Grade 8, there was considerable emphasis on algebra. The work, however, was almost entirely based on abstract manipulation of symbols and I noted no examples of contextualised "situations".

Elsewhere, the negative integers were introduced at the end of a content-packed chapter which also included a definition of positive whole numbers, the number line for these, and the concept of ordering, together with the signs for greater and less than, work on primes and factorisation, and rounding. Again, negative numbers were introduced via the thermometer and emphasis placed on ordering integers, before addition and subtraction were introduced by means of vectors operating on the number line. Rules and definitions were clearly set out and numerous formal exercises given in addition to the occasional contextualised one. As in several countries, multiplication was approached via repeated addition when one factor is positive. The pattern $(-3)\times 3 = -9$, $(-3)\times 2 = -6$, ... was then used to justify the statement that $(-a)\times(-b) = ab$. The chapter ended with the rules for division of negative numbers and many further abstract exercises.

A seventh country attempted to solve the problem of differentiation through the use of extremely long textbooks from which teachers selected a course appropriate for particular types of pupils. As a result, chapters entitled "Integers and Rational Numbers" appeared, with slight differences, in both the Grade 7 and Grade 8 texts. The latter began with a novel contextualised example of negative numbers based on the daily changes in share prices being given in the financial press using numbers such as +1 and -5/8. (This again raises the problem of using "realistic" contexts that lie outside the experience of the vast majority of students.) After that, the mixture was fairly standard: heights and depths, the number line, opposites, absolute values and ordering. There was a good mix of questions including the translation of literary phrases into mathematical language and some more taxing questions such as: Is there a least negative integer? A greatest negative integer? Again, the addition and subtraction of integers was introduced via shifts in temperature and on the number line, and multiplication through the completion of patterns. The calculator was then used to check multiplication of integers and the kernel for multiplication stated. Amid the formal exercises were items such as, "What is the product of three negative integers?" and "State a rule for predicting the sign of a negative number raised to an even or odd power". Again, I did not spot any exercise in which multiplication of integers was set in context.

The final country has a stratified system and two sets of texts were used at this level. The higher-ability students had already dealt with operations on the integers and in Grade 8 plotted the graphs of functions such as $y = x^3 - x$ and solved quadratic equations. The lower-ability students were presented with an approach to the integers that I am told is traditional in that country. This approach attempts to provide meaning through context, in this case a fable. The seeds of negative numbers had been sown in Grade 7, and preliminary work done on addition and subtraction. The first of the two chapters on integers in the Grade 8 book took up the story from there. We were told of a witch who possessed two special kinds of cube. One, let us call it black, would, if placed in her cauldron, raise the temperature of its contents by one degree. The other, red, lowered the temperature by one degree. The addition of cubes was treated as equivalent to vector shifts on the number line. This permitted the definition of addition (note the similarities with the mathematicians' "ordered pair" definition of the negative numbers), after which other examples were introduced, such as drawing money from a bank, and use made of calculators. Subtraction depended upon the witch's ability to extract cubes from her cauldron with the desired mathematical effect. This led to the needed rule, and the chapter effectively closed with some interesting and far from trivial work on coordinates and graphing.

Multiplication and division were treated in a later chapter of the same text. Now the witch threw packets of, say, 5 red cubes into her cauldron and observed the results. Mathematically, we have the analogue of the vector approach; that is, the cubes (or packets of cubes) represent vectors, and effectively a scalar multiple of a vector is being defined. By ignoring the difference between the scalar and the vector, that is, between 3 representing a number of packets and 3 representing three black cubes, we obtain the required rule for multiplication. (Any attempt to mirror the "ordered pair" approach to multiplication would, of course, have been impossible.) Again, packets of cubes can be extracted and the laws of thermodynamics ignored. Many exercises were provided within the witch's cauldron context before abstract formal exercises were introduced. Some exercises using coordinates were provided before division was introduced in a more formal fashion.

These examples on the introduction of operations on the integers repay study because they lie astride the "commonsense" boundary. In the countries from which these texts were taken, negative numbers, but not operations on them, are now part of common sense (i.e., of local knowledge), hence their appearance in the primary school grades. Weather forecasters tell us that temperatures will rise to x, or fall to y, rather than that they will rise by u degrees or fall by v; yet, connections are so close that the introduction of addition and subtraction do not seem to cause great difficulty. Looking again at Vergnaud's definition of a conceptual field, we see that there is a variety of situations in which one can apply the new concepts, and that these are firmly associated with ideas, and with meaning, which students will bring to the lessons. Attempts to explain addition and subtraction through the use of vector shifts on the number line may give rise to some confusion, since, say, the sign -1 will be used to denote both a milestone on the number line *and* a vector. Nevertheless, the association of *meaning,* in all its senses, would not seem too difficult in relation to the introduction of negative numbers and of the operations of addition and subtraction.

Greater problems arise with the multiplication of integers. Is this going to be "defined", or are attempts to be made to establish, through the use of common sense, a "theorem"—with plausibility replacing proof? Are the results to be restricted to the integers (in which case repeated addition can to some extent be used as a means of justification), or do we wish to deal with the reals? The problems of legitimation in the latter case are, of course, immense. To what extent do we wish to make explicit the historical thought processes that led to the definition of the multiplication of integers? Some years ago, the poet, W. H. Auden, explained how at school he was taught: "Minus times minus equals plus. The reason for this we need not discuss".

How does one improve on this approach? What can be done to supply that "set of situations" that will give meaning to this new concept? How does one explain why mathematicians were led to extend the definition of these two operations to the integers?

Certainly, the texts I read did not provide very satisfactory solutions to these problems. No convincing metaphors were to be found, and in pedagogical terms there was a catastrophic jump from the previously largely commonsense-based, and often contextually rich, approach to topics to a feebly motivated, abstract treatment of the multiplication of integers.

A close identification of negative integers with physical measurements or quantities, be they temperatures, debts, or vector shifts, will only make the association of meaning with multiplication more difficult (or be achieved through mathematical sleight of hand). Working on a common-sense contextualised level, while helping with the introduction of addition, will then, unless attempts are specifically made to move to a more abstract view, only store up potential trouble. But what is to be the *motivation* for studying negative numbers and their multiplication? Historically, the major impetus has been the need to extend algebraic methods. (A brief sketch of ideas concerning negative numbers from Brahmagupta, through the puzzled sixteenth and seventeenth century mathematicians, to the agnostics of the eighteenth and nineteenth centuries, can be found in Kline, 1972.) It would still seem important that this motivation be explicitly stated, or implicitly assumed through practice, in any modern approach. Yet I can recall no mention of the need to extend the number system *and operations on it* to allow us to carry out formal operations, for example, to give meaning to the "solution" of previously insoluble equations. Indeed, no text introduced these operations in order to *do* anything: motivation seemed entirely lacking. The content was in the syllabus and that was sufficient justification. Of course, finding contextualised examples requiring the multiplication of negative numbers is not easy, but more could have been done. After reading the various texts, I turned to a book written in 1931 by the English schoolteacher, C. V. Durell, *The Teaching of Elementary Algebra*. Let me summarise what I found.

Durell begins by stressing history as strong evidence of the difficulty of the concepts involved, and remarks that the introduction of directed numbers had, in recent years, tended to be delayed as a result of the underlying theory and difficulties being better appreciated. He, himself, strongly believed that "it is inexcusably wrong to allow or to teach pupils to use symbols to which they attach no meaning". One consequence of this was his wish clearly to distinguish between a signed negative number and the operation of subtraction on unsigned numbers. The approach he suggests for the introduction of directed numbers and for addition and subtraction are those most favoured today: temperatures, gain and loss, and so on, but with an explicit emphasis on one-dimensional movement. Here the emphasis should be placed on the fact that "the 'rules of signs' are definitions that have been framed to establish correspondences between similar processes in different number systems. The question of *proof* does not arise here, though it does so in connection with the consequences of the rules, but [such work] is for specialists". On multiplication, he suggests beginning with consideration of repeated addition, e.g., $((-5)\times 3$. "This does not *prove* that $(-5)\times 3 = 15$; it would merely make matters very awkward if it were not so, and suggests what kind of definition is most useful". Further justification for a definition is sought through contextualised examples. Here we want to substitute negative numbers into physical formulas that involve products.

The number of these available to Grade 8 students is limited! However, by using time measured from a fixed hour, we can employ the simpler equations arising in kinematics, or, another of his examples, temperatures rising at a steady rate in a boiler. Durell asks students to supply interpretations of these formulas when none, one, or two of the variables have negative values. And so his advice proceeds.

A frightening aspect of all this is that I find Durell's approach more pedagogically and mathematically appealing than any to be found in the textbooks I studied. Durell was an exceptional teacher; his pupils included mathematicians of the stature of Freeman Dyson and James Lighthill, but what he suggests is capable of ready adoption by less gifted teachers. How is it that we fail to build upon the legacy of the past? Since 1931 "mathematics education" has grown into a profession and, possibly, a discipline. Yet how much closer are we to finding an effective and mathematically truthful way of introducing this key concept to be found in the curricula of every country? Have emphases within mathematics education been sensibly placed? Is modern technology being used to best advantage in a mathematically and pedagogically thoughtful way?

But let us not end on a discouraging note. Elsewhere in the texts I studied, I found an altogether more encouraging example: a new approach to the problem of introducing formal proof that made use of the concept of flow diagrams. One cannot tell from a textbook how successful a method will prove in a classroom, but what was readily apparent was the thought underpinning this innovation. Clearly such developmental research work—uniting both mathematical and pedagogical thought—is only one aspect of mathematics education. Yet, if we are to succeed in ensuring that students attach more meaning to mathematics, in all the senses described earlier, it is essential that mathematics educators give this work a higher priority than it has received in recent years.

REFERENCES

Du Bois-Reymond, P. (1882). *Die allgemeine Funktionentheorie* (quoted in translation in Kline, 1972, p. 998).

Durell, C. V. (1931). *The teaching of elementary algebra.* London: Bell.

Freudenthal, H. (1991). *Revisiting mathematics education.* Dordrecht: Kluwer.

Howson, A. G. (1995). *Mathematics textbooks: A comparative study of grade 8 texts.* Vancouver: Pacific Educational Press.

Howson, A. G., Harries, T. & Sutherland, R. (1999). *Primary school mathematics textbooks.* London: Qualifications and Curriculum Authority.

Kline, M. (1972). *Mathematical thought from ancient to modern times.* Oxford: Oxford University Press.

Otte, M. (1986). What is a text? In B. Christiansen, A. G. Howson & M. Otte (Eds.), *Perspectives on mathematics education* (pp. 173–204). Dordrecht: Reidel.

Rowe, D. E. (1994). The philosophical views of Klein and Hilbert. In S. Chikara et al. (Eds.), *The intersection of history and mathematics* (pp. 187–202). Boston: Birkhäuser.

Sierpinska, A. (2000). On some aspects of students' thinking in linear algebra. In J.-L. Dorier (Ed.), *On the teaching of linear algebra* (pp. 209–246). Dordrecht: Kluwer.

Thom, R. (1973). Modern mathematics: Does it exist? In A. G. Howson (Ed.), *Developments in mathematical education* (pp. 194–209). Cambridge: Cambridge University Press.

Vergnaud, G. (1983). Multiplicative structures. In R. Lesh & M. Landau (Eds.), *Acquisition of mathematics concepts and processes* (pp. 127–174). New York: Academic Press.

MARIA G. BARTOLINI BUSSI

THE MEANING OF CONICS: HISTORICAL AND DIDACTICAL DIMENSIONS

In this paper, I draw on an analysis of the research project *Mathematical Machines,* which concerns the teaching and learning of geometry in high school (grades 9 to 13). Although the project is actually broader (see Bartolini Bussi, 1993, 1998, 2000, 2001; Bartolini Bussi & Mariotti, 1999; Bartolini Bussi et al., 1999; Bartolini Bussi & Pergola, 1994, 1996), I have chosen the special topic of *conic sections* (or *conics*), which I take to be representative of the whole approach.

My main thesis is that *the present meaning of conics is the result of the complex relationships between the different processes of studying conics during different historical ages*, each of which has left a residue in the names, the problems, the means of representations, the rules of actions, and the systems of control. To investigate the present meaning, we may refer to the historical development of the study of conics by means of *time periods*, each framed in the culture of a different age. Even if from today's standpoint all the conics studied in the different time periods can be identified as the same objects, inside each time period different meanings have been built by geometers to the extent that conics are representative of the development of different conceptualisations of space and geometry over the ages. As a corollary, I claim that *it is not possible to build the meaning of conics through only a one-sided approach*, as, for instance, through the most widespread algebraic definition.

If history is an unavoidable component of the construction of meaning, a didactic problem immediately arises: *How is it possible to introduce students to the historical problematic without undue oversimplification?*

An exemplary teaching experiment will be described to show how the problem of *epistemological complexity* (as meant by Hanna & Jahnke, 1994), on the one hand, and the problem of *historical contextualisation,* on the other, are coped with by means of a selection of tasks. Finally, a small-group study of a special model for the parabola (the *orthotome,* inherited from the Greek tradition) will be analysed. It concerns how the meaning is constructed by students through the introduction of a conscious anachronism that fosters an intentional recourse to different tools developed in different ages and allows the students to relate different ways of representation to each other.

1. THE MEANING OF CONICS: HISTORICAL DIMENSION

Consider the special topic of *conics* (for centuries called *conic sections* to emphasise the generation by means of a cone). I claim that the most common approach, which is through analytic geometry (i.e., conics as plane loci satisfying equations of the second degree obtained from some metric relations), is not enough. A lot of the meaning of conics is lost: Where do their names come from? Why are they studied together? Why do they have some special importance in geometry? And so on.

The algebraic approach is deepened, if, as happens in university courses, geometry is replaced by linear algebra and more notions are added. Quadratic forms in three variables are considered as special cases of quadratic forms in $n + 1$ variables that define hyperquadrics in n-dimensional complex projective space: Terms like *cone, cylinder, diameter, axis,* and so on are used. Why? Bourbaki (1974) claims, in his historical reconstruction of the genesis of quadratic forms, that in the search for greater and greater "abstraction", it has been considered very suggestive and attractive to preserve the terminology that originated in the study of cases of two and three dimensions from classical geometry and to extend it to the case of n dimensions, to the extent that geometry has been transformed into an universal language for contemporary mathematics. But surely the intention to convey suggestion and attraction can be realised only for people who also know the spatial generation of conics.

Asked the question, What is the meaning of conics? (which is related to the question of the meaning of quadratic forms), one can give many answers. To cite just a few, one can consider conics analytically as curves of second degree, synthetically in three-dimensional space as conic sections, in the plane as loci satisfying some metric conditions, as perceived images of a circle from a variable point of view, and so on. All these interpretations are related to each other, yet they are concerned with different conceptualisation of conics that can be related now by means of the existing body of knowledge.

To explain the above discourse, it is necessary to reconstruct the historical development of conics. Obviously, the figural representations of conics (as signs traced by means of a gesture, linkage, or cut made either in the sand, on paper, in the air, or on the surface of a cone, and so on) are invariant in time and hence not subject to historical changes. What are changed are the way of generating conics, the way of looking at them, and the way of studying them.

These attitudes are actually more general. The history of conics (as well as the history of any mathematical object that dates back to antiquity and is still part of today's mathematical culture) is a metonymy for a more general history, namely, the history of the geometrical conceptualisation of space. As such, it cannot be understood inside mathematics only and requires references to the complex relationships between mathematics and general culture.

In short (for details, see Coolidge, 1945), one may identify four major phases:

1. Greek mathematics, where the early emergence of conic sections is documented.
2. The 17th century.

3. The 18th and 19th centuries.
4. The 20th century.

The two initial phases are of critical importance: The first concerns the birth of conics as geometric objects; the second concerns the emergence of the trends of discussion that characterise the modern treatment of conics. The jump from Greek mathematics to the 17th century is due not to a naive underestimation of the contributions of the Middle Ages and the Renaissance (they actually constitute the ground on which subsequent development is based) but to the fact that in the 17th century one encounters the early results of a complex social phenomenon that radically changed the attitude towards mathematics, the sciences, and technology—that is, the interaction of merchants, scientists, engineers, artists, medical practitioners, humanists, and so on, and its amplification through the increasingly widespread diffusion of ideas by means of the printing press (Otte, 1993). The eruption of new ideas in mathematics is visible also in the approach to conics, where new conceptual tools were being accepted from outside mathematics: for example, from commercial arithmetic (i.e., algebra), from the arts (i.e., the introduction of points at infinity in perspective drawing and the study of anamorphoses in painting), and from technology (i.e., machines for drawing curves).

In the next two centuries, this new attitude was developed and carried to extremes: The complete algebraisation of conics allowed the development of the theory of quadratic forms in connection with problems from arithmetic, analysis, and mechanics; the great projective school allowed the characterisation of conics on the basis of location and intersection rather than of metric properties; the theory of articulated systems, developed in connection with kinematics, on the one hand, and the theoretical treatment of geometric transformations, on the other, allowed the characterisation of algebraic curves (which include conics) as the curves that can be traced by linkages. Later, in the 20th century, the Bourbakist program for the complete algebraisation of conics up to the theory of quadratic forms had a great effect on devisualisation, and only in the last decade of the century did the introduction of computer aids reintroduce a visual dimension into that purely algebraic world.

Within each period of time, different objects were built by geometers. What the objects have in common is the name (and sometimes not even that, as we shall see) and some classical problems. The difference is so deep that mathematicians often feel obliged *to prove* that the new objects are actually the same as in the past.

There is not space in this chapter to offer an account of the different time periods. For the purpose of this paper, I limit myself to reminding the reader that at the beginning of the story, a parabola was obtained by Greek geometers by cutting a right-angled right circular cone with a plane perpendicular to one element of the cone (whence the Greek name *orthotome*), whilst the curves today known as ellipse and hyperbola were obtained by cutting an acute-angled and an obtuse-angled cone (whence the Greek names of *oxytome* and *amblytome*) in the same way (some details on these issues are in Coolidge, 1945; a proof in the case of orthotome is given in the section "Students' construction of meaning" below).

Later, things changed. For instance, Descartes (Bos, 1981) treated the curves in totally different ways according to their role as either a solution of a problem (a product) or a means of finding solutions. A pointwise construction of a conic by ruler and compass is sufficient when it occurs as a solution, but when in a problem it is necessary to find solutions by intersecting conics, a pointwise construction is no longer sufficient. A stronger criterion is required: It is necessary to have a method to trace the curve by means of "some regular motion" that allows one to find *all* the points so that the intersections with other lines are precisely (and not only approximately) constructable. The problem of continuity is solved by referring to motion and time. In the same fashion, further developments added new elements to the meaning of conics to constitute a complex object.

2. THE MEANING OF CONICS: DIDACTIC DIMENSION

The present meaning of conics is the result of an accretion of terms, problems, ways of representations, rules of actions and systems of control that have been inherited from the different time periods. This fact has important consequences on the didactic plane: To construct the meaning of conics in the classroom, it is necessary to reconstruct some elements of their historical development.

This is actually a didactic problem: Is it possible to introduce students to the historical problematic without undue oversimplification? In what way? This section of the paper is devoted to that issue, showing how my colleagues and I designed and implemented a field of experience for students' activity in the classroom in order to implement the historical reconstruction of the meaning of conics. The experiments were carried on in secondary classrooms up to the late 1990s. Later, they were shifted to the university level and to pre-service teacher education.

The context of classroom activity is characterised by the presence of physical models (either static or dynamic), which are not simply shown to students but are objects for students' investigation. So, for instance, students are given a three-dimensional model of a conic section or a plane trammel that draws a conic, and the task is to determine the geometrical properties of the points on the curve. These models are historically contextualised by means of guided reading of historical sources.

I illustrate this process by means of three kinds of data:

a) a scheme of a long-term teaching experiment designed and implemented in the classroom to realise the main motives of the whole activity,
b) the analysis of a task whose goals are consistent with the motives of the whole activity,
c) the analysis of a small-group session up to the product of a collective written text.

2.1 Motives of the teaching-learning activity for Mathematical Machines

The teaching-learning activity in the research project *Mathematical Machines* is polimotivated. The term *motives* is used after Leont'ev (1978) to mean the objects of an *activity*, to be distinguished from *goals* (or aims) and *conditions*. The macrostructure of an activity consists of *actions*, each of which is directed to a specific goal; the ways of realising actions in concrete conditions are *operations*. The study of long-term processes concerns the relationships between the levels of teacher's motives, actions, and operations and the effects produced on students' activity.

The motives can be briefly sketched as follows (they are surely mutually intertwined):

1. the conceptualisation of mathematics not as an isolated body of knowledge but as a part of the global cultural development of mankind to be studied in its relationships with other fields of knowledge,
2. the historical contextualisation of accepted rules of behaviour,
3. the multifaceted meaning of conics aside from the more usual plane meaning as loci of points defined by metric relations,
4. the dynamic interpretation of either dynamical or static objects used to guess conjectures and to guide the construction of early proofs, plus the introduction of movements and of the principle of continuity (implicitly) to cope with the problem of "generic" points

The motives can be determined a priori by the analysis of the teacher's programmatic documents (Pergola & Zanoli, 1994, 1995) and can be checked a posteriori by either the definition of school tasks or the quality of interaction in the classroom. Below, I give the scheme of the tasks of a particular teaching experiment and some exemplary excerpts of small-group interaction during the study of the orthotome.

2.2 Tasks in the laboratory of Mathematical Machines

Student activity takes place in a special room (the mathematical laboratory), where several physical models (either static or dynamic) are at the students' disposal (two catalogues of the models can be found on line at http://www.mmlab.unimore.it). Large-sized models (built on bases that are more than 60 cm by 60 cm) have been built by the teachers themselves using wood, brass, plexiglas, coloured threads, sinkers, and so on. Models are sometimes used by the teacher to illustrate a concept, but more often they are handled by the students themselves in order to examine them according to some specific task. Concrete handling of models is contextualised by means of the guided reading of some selected and annotated historical sources.

For the special topic under scrutiny, several models are available for either solid or plane study and for either static contemplation or the dynamic generation of conics. We have models from the Greek period (e.g., models from Menaechmus and Apollonius) and models from the 16th to 20th centuries (e.g., models for the mechanical generation and the projective study of conics). However, the models alone are opaque; only the reading of sources and guided manipulation can make the different conceptualisations explicit.

Figure 1: The model of the orthotome studied by students in small-group work

2.3 A teaching experiment

Long-term teaching experiments are designed and implemented in the laboratory as standard part of the curriculum. In some cases, the model substitutes for the need for an explicit proof; in other cases, it generates the need for a proof. In any case, it allows the teacher to introduce historical digressions that contextualise the study in the culture of the corresponding age. The first part of an exemplary teaching experiment is described below (see Tufo, 1995).

2.3.1 The teacher's lesson: A historical introduction to conics in ancient Greece

The teacher motivates the historical introduction with the need to reconstruct a deep geometrical meaning for the algebraic relationships between coordinates (already known by students and applied to a classical derivation of conics from focal properties) that represent conics in the Cartesian plane. He illustrates the conceptual difference between the way of looking at conic sections in ancient Greece and at conics in the modern age (since the 16th century). He focuses on three issues:

1. conics as solid curves versus conics as plane curves,
2. conics as existing objects to be studied (plane sections of cones) versus conics as products (representations of laws or drawings by instruments),
3. synthetic versus analytic study of conics

2.3.2 Small-group work: The study of the orthotome model

The whole class is divided into five small groups, each studying a different model. A small group of four students is given the task of studying the model of the *orthotome* and of deriving the "symptom" of the *parabola*. The students are prompted to refer to theorems on right triangles and on similar right triangles and to avoid recourse to coordinates, as coordinates and analytic geometry did not exist in ancient Greece. They have to state the symptom and to prove it, in order to be able to explain it to their schoolmates in a later lesson. Small-group work is carried on with some interventions by the teacher, who walks around the classroom to observe the groups at work. A final written report is requested from the group (see below).

2.3.3 Students' explanation of the orthotome model

Two students in the group present to the whole class the result and the proof arrived at in their small-group work.

2.3.4 The teacher's lesson: Models of the orthotome and the equations of conics

The model of the orthotome is considered again: A system of coordinates is introduced to derive the canonical equation of the parabola

$$2\,k\,x = y^2$$

2.3.5 The rest of the experiment

A leap is made to the 17th century to introduce the study of conics according to de l'Hospital. The teacher introduces this different approach by means of conic-drawing instruments, discussing the changes in attitudes towards geometry. Then he proposes the study of conics by means of three different definitions from de l'Hospital, based on conic-drawing instruments. They are quite different from each other and make it clear than the conceptualisation of ellipse, hyperbola, and parabola as different manifestations of the same geometrical object is quite difficult from a metric perspective.

2.3.6 Students' construction of meaning

In this section, I discuss the small-group work up to the collective essay produced by the students at the end of their study of the orthotome model. The final text was produced collectively after a 2-hour small-group laboratory on the model of the orthotome. In this section, I use a two-column format: In the right column are transcripts of discussions and texts produced by the students; in the left column are the students' original drawings and comments by my colleagues and me.

2.3.7 The quality of small group interaction

The task, assigned verbally by the teacher, was the following (according to the complete transcript of the laboratory; see Tufo, 1995):

Teacher: You have to obtain an important property of parabola. I can help you a bit: It is the property that Greek geometers obtained by examining this situation, where the parabola is already drawn. As you see, it is in three-dimensional space, on the surface of the cone. It is the same one that is described by the Cartesian equation of the parabola that you know. You have to discover the relationship between the green line segment [VS in Figure 2] and this line segment [PS in Figure 2].

1. The task
In this task, the conjecture phase is cut short: The students already know the property, which is expressed by the usual canonical form of the parabola equation. In other tasks (see Bartolini Bussi, 1993), the conjecture phase is explicitly assumed as a part of the task.

The teacher clearly states that the study in the solid setting and in the algebraic setting can be considered "the same": it is an example of regressive appropriation that depends on today's knowledge.

The small-group work can be divided into several episodes, which we have numbered and labelled. Some of these episodes contain joint activity with the teacher. Below are some exemplary short excerpts that are related to different issues:

a) the quality of help offered by the teacher to introduce the problem,
b) the quality of help offered by the teacher to explore the model,
c) the quality of dynamic exploration carried out by the students,
d) the quality of help offered by the teacher to sum up the whole process.

The excerpts have been chosen from the complete transcript to illustrate some critical features of joint activity. In particular, the issues (b) and (c) are related, as they represent some of the teacher's operations and the process thereby induced in the students. The effects of the help offered in (a) and (d) is better acknowledged in the final written report, which is analysed in the next section.

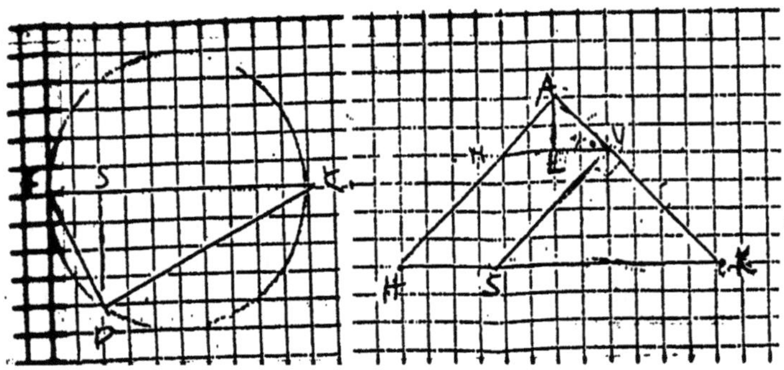

Figure 2: Figures produced by students during group work with coding letters

T: [...] The problem is to discover, on the basis of well-known theorems, theorems about right angles and similar right angles, which is the property that links these line

2. The teacher's introduction
In this long introduction, the teacher offers help by referring to some geometrical figures (right angles) to be focused on. He also

segments, the green segment that we call abscissa and this segment that we call ordinate. You have to observe the model, this circle, this right-angled cone. You have to elicit all these hypothetical elements; then, by reasoning about them, you have to obtain a property of parabola that is perfectly equivalent to the equation that we write now, to the equation that Descartes would write many centuries later. It is the same relation, and it is possible to obtain it only in space. Look and try, say, for a quarter on an hour. It's a reasonable time. Then if you do not have any idea, call me; otherwise, go on with the things you have seen.

[...]

defines the degrees of freedom for the students: They cannot work for days but have a limited time to explore the model. At the same time, he introduces all the motives:

- the reference to history, to justify the accepted rules of behaviour
- the equivalence of synthetic and analytic description and their contribution to the meaning of parabola
- the necessary reference to the physical object so as to guess conjectures and be guided in proving

These elements will be refocused again and again during the small group to construct the sense of activity.

The students start to draw the fundamental elements of the model.

T: Sometimes it is necessary to consider also figures that are not immediately visible. For instance, in this case there is a right triangle that is fundamental, but it is not traced yet. You can look for it by taking into account that you need to consider those triangles whose sides are among the lines segments you have to relate to each other.
S 1: There is a right triangle [*OPS*]... It changes; when you change the plane the base is changed...
S 2: Yes, but in the meantime this other triangle changes...
T: This exploration seems a good idea. You have to reason in plane geometry but not always in the same plane; remember that Greek geometers saw the figure in space, even when they did plane reasoning, by considering different planes

[...]

3. Helping to explore
The teacher also gives methodological help to students to direct their search towards effective strategies. Students put into practice the teacher's suggestions: They have a rigid model, made of wood, Plexiglas, and thread; they cannot do real experiments (as in Cabri with dragging), but they see the changes. (The letters in brackets are only for the reader's help; see Figure 2.)

In this phase, the recourse to coding points by letters is very limited; the teacher too points at the model and speaks about "this triangle" and "this point". Coding with letters becomes essential (to the problem) only later, when proportions are to be written down. In this phase, it is more useful to build a dynamic ideal object on which to make a mental experiment for guessing conjectures.

S 2: Yes, this [$Y = PS$] is the height of the right triangle [*PHK*], *PS* the point [sic!]. *PS* is the height of the right triangle that is going to be formed [Italian: si viene a formare] in the semicircle. The angle in the semicircle is always 90°, isn't it?
S 1: It is! Hence this triangle...
S 3: Practically there is the triangle that

4. The dynamic exploration
The habit of moving a static object is typical of this classroom. I have explicitly stressed an unusual Italian expression (*si viene a formare*) that emphasises the progressive formation of an object that does not yet exist in the figure but is created at this moment in the mental process. Surely not all the

rotates on the semicircle; that is, it moves on the semicircle.
S 2: Only the height of this triangle is changing [she gestures with her hand palm upwards to show that when the horizontal plane is going up the height of the triangle *PHK* is changing].
S 3: Considering the two planes [i.e., the horizontal and the oblique plane]... Yes, but there is also this plane [the vertical plane].
S 2: In this plane [vertical], if we consider this triangle [*VAS*] when the plane changes, when the plane moves, it changes too.

explorations are effective for the solution; they are indeed necessary to create the dynamic ideal object.

The process of building a proof is long. It is necessary to choose the useful triangles and to mark on them the useful proportions. The proportions are to be interpreted within the theory of application of areas (*to behave like Greek geometers*), up to the statement of the *symptom* of parabola:

$$2\ VA\ VK = PS^2$$

Only later does the teacher suggest relating this formula to the post–Cartesian approach. It is done by making the following substitutions:

$$VK = VS = x \qquad PS = y \qquad VA = k$$

This yields:

$$2\,k\,x = y^2$$

The final interpretation is done by the teacher, who sums up all that has been already said during the group work:

T: When we write

$$2\ VA\ VK = PS^2$$

or

$$2\,k\,x = y^2$$

it is the same. Only the notations are changed. This is the modern notation, and that is the one that they [the Greeks] used, but the equation is the same: It is the equation of the parabola. It is the geometric property of parabola.

That [i.e., *VA*] is constant, since when I move this one [he gestures to indicate moving up and down the horizontal plane], this does not change, because the vertex of the cone and the secant [oblique] plane are

5. Helping to sum up the process
At first, the relation between the proportion and the equation is recalled. This is a way to emphasise the spatial interpretation of the equation of the parabola that otherwise risks being lost.

The introduction of the constant k is justified on the model by observing that a part of the configuration does not change when the auxiliary horizontal plane is changed.

The change in the secant plane is *imagined* on the fixed model by gesturing.

the same.

If I cut with a plane closer to the vertex, how does the parabola change?

Because, if I keep the secant plane fixed and change this [the horizontal] plane, the parabola is the same; what changes is only this arch that becomes longer or shorter. But if I change the secant plane, and I cut the cone with a secant plane closer to the vertex of parabola, how does the parabola change?

Ss: [looking at the model] It narrows.

T: Right, it narrows. And if this becomes longer, if this plane goes here...

Ss: It widens.

T: Right. And this is what happens [he points at the equation], isn't it?

Let the equation of the parabola be

$$x = h\, y^2$$

If this [h] changes, the width of the parabola changes.

The secant plane goes closer to the vertex...

...or farther from the vertex...

...and this change is related to the equation.

2.3.8 The final report

The report is produced as a collective homework by the group after the laboratory ends, on the basis of drawings and personal notes. In the left column below, the students' text (before any correction by the teacher) is translated literally. In the right column, a division into sections is suggested. This division clearly shows that the students have always succeeded in giving a logical organisation to the long text. The complete analysis of the transcript (Tufo, 1995) shows that the order is not the same order of the exploration during the group work: Then it represented a further control on the collective activity.

Plane Section of a Right-Angled Cone

It is necessary to distinguish between RIGHT cone and RIGHT-ANGLED cone. On the one hand, a right cone is when the perpendicular from the vertex to the plane of the circle (i.e., the directrix of the cone) is in the centre of the circle itself. On the other hand, a right-angled cone is generated by rotating an isosceles right-angled triangle about a cathetus [leg]. The cone is called right-angled as the angle between two opposite generatrices is 90 degrees. If it were acute, the cone would be acute-angled; if obtuse, obtuse-angled.

1. Definition of right-angled right cone

The students start from a possible misunderstanding about right and right-angled cone. They probably remember a personal experience. They recall Euclid's definition and relate the definition of right-angled cone to other kinds of cones. This means relating the orthotome to other conic sections.

They are consciously in the solid setting and are consciously using an approach inspired by history. A few lines below, they make an explicit reference.

Description
The model reproduces a RIGHT-ANGLED cone generated through the rotation of an isosceles right triangle cut by a plane perpendicular to the generatrix *AH* and parallel to the opposite generatrix *AK*. The model contains also two perpendicular PLANES: the plane *t* of the circle (directrix of the cone) and the plane *t'* (MERIDIAN) of the axis *AO* of the cone. The plane *t'* contains the segment *VS*, from the vertex of the section (ORTHOTOME) to the point *S* (the intersection of the diameter *HK* of the circle of the plane *t* with the line *PM* of the secant plane).

The section of a right-angled cone produces a curve named ORTHOTOME by the ancient geometers prior to Apollonius.

2. Description of the model
The physical object is carefully described. Two planes are explicitly named: The former [*t*] is a physical plane, realised by a wooden base; the latter [*t'*] is an ideal plane, determined by a wooden frame. The secant plane is made of plexiglas.

The reference to history is explicit.

The property or SYMPTOM (verified by all the points of the curve), that allows one to recognise the kind of plane section of the cone is based on the ~~equality~~ [equality is erased] relationship between the segments *PS* and *VS*, i.e., on the ~~equality~~ [equality is erased] equivalence of two geometrical figures that individualise the position of the point P.

3. Task
The students are recalling the task and the accepted rules of behaviour that have been stated by the teacher in the contract. They have to behave like Greek geometers and use proportions and equivalence of areas. An incorrect term is erased by them and replaced by a better one.

The reasoning from which the property is drawn, a property valid not for a special point of the orthotome, as the secant plane maintains always the same distance from the vertex of the cone and the same slope, is the following:

(Generalisation to any point of the orthotome)
Someone might believe that the reasoning works only for the special point of the figure: The students explain that it works for every point, anticipating now what will be argued at the end (see point 9 below).

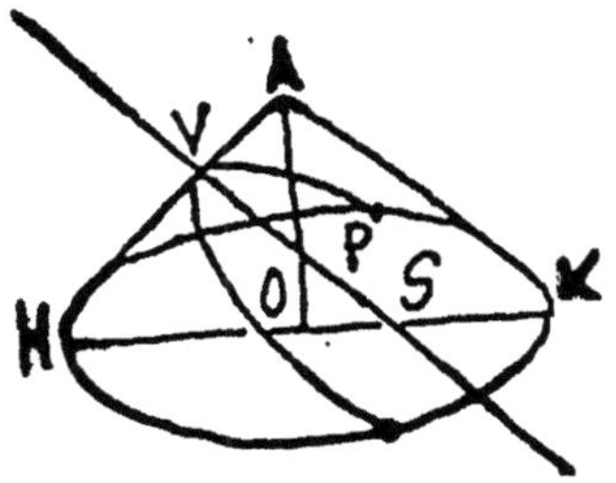

Figure 3: The first drawing produced by the students

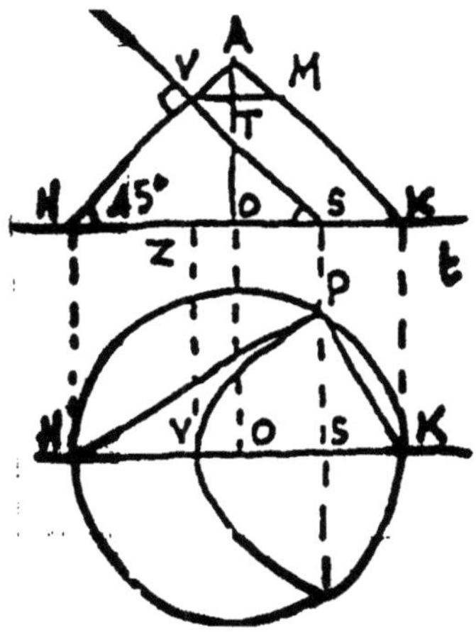

Figure 4: The second drawing produced by the students

Let us apply the reasoning first to the plane t on which the circle is.

If we consider point *P* of the orthotome, we observe that it, obtained from the intersection between the secant plane and the directrix of the cone, lies on the circle, whatever the distance *AV* between the vertex of the cone and the plane. Every triangle inscribed in a semicircle has an angle of 90°; hence the triangle *HPK* inscribed in the semicircle with diameter *HK* is right-angled, with HPK = 90° (THEOREM). Hence it is possible to apply the *theorem of Euclid* in

4. Reasoning on the plane *t*

The students are using the theorem of Euclid for right triangles.

They state the theorem referring to mean proportionals...

right-angled triangles. The height *PS*, drawn by means of the perpendicular from *P* to the diameter *HK*, is the mean proportional between the projections of the catheti [legs] of the triangle on the hypotenuse, i.e. the diameter *HK*.

We state the proportion ($HPK = 90°$):

$$HS : PS = PS : SK$$

that is,

$$PS^2 = HS \cdot SK$$

We have proved that the area of the square built on the segment *PS* is equal to the area of the rectangle with sides *KS* and *HS* (geometrical interpretation).

...and interpret the proportion, like Greek geometers, as an equivalence of areas.

Let us consider now the meridian plane *t'* perpendicular to the former one. We try to state a proportion that can relate *PS* and *VS*. We observe that on the meridian plane there are two similar triangles. The former is *HVS*, with hypotenuse the projection of the cathetus *PH* on the diameter *HK* and cathetus the distance from the vertex of the orthotome to the intersection of the generatrix *AH* with the circle. The two catheti *VS* and *VH* are equal, as the angle *VSZ* is 45° (the secant plane is perpendicular to the generatrix *AH*, and the angle *VHZ* is 45° (*AHO* is an isosceles right triangle.

Hence the triangles *VZH* and *VZS* are equal, whence $VS = VH$.

The other right triangle to be considered is *AVT*, formed by the line *VM* parallel to the diameter *HK* and by the axis *AO* of the cone.

As the angles $TAM = VAT = 45°$, $HAK = 90°$. As they both have an angle of 90°, $AVT = AMT = 45°$ and $VT = AT$.

The triangle *AVT*, as an isosceles right triangle, is similar to the triangle *HVS*. We can state the proportions:

$$HS : VH = AV : VT$$

hypotenuse : cathetus = hypotenuse : cathetus

That is,

5. Reasoning on the plane *t'*

Among the many triangles and proportions that could be observed in the model, the students claim to focus on a relationship between *PS* and *VS*. This statement clearly represents a conscious control of the strategy. In the workgroup, explorations have been actually much more extensive: not all the explorations have proved to be useful.

The two isosceles triangles are found. The proportion is stated with the control of the meaning...

...and translated into an equivalence of areas.

$$AV \cdot VH = HS \cdot VT$$

If we compare the proportion with the above relationship

$$PS \cdot PS = HS \cdot SK$$

we observe that

$$HS = 2 \cdot VT$$

as $VS \parallel MK$, i.e., $VMSK$ is a parallelogram and the triangles

$$AVT = ATM.$$

We obtain:

$$SK = 2 \cdot VT$$

$$PS \cdot PS = HS \cdot SK.$$

As in both the first and the second equations [sic], there are both HS and SK

$$2\,AV \cdot VH = 2\,HS \cdot VT$$

multiplying each member by two in order to obtain

$$2\,VT = SK$$

whence

$$PS^2 = 2\,AV \cdot VH.$$

The area of the square built on the height PS is equal to twice the area of the rectangle with sides $VH = VS$ and AV.

6. Linking the two planes together
Up to now, two different steps have been realised: the former in the base plane t, the latter in the meridian plane t'. A link between them can be found.

A fundamental relation is obtained (the *symptom*) and interpreted as equivalence of areas.

If we introduce a suitable system of coordinates x and y in the secant plane, the coordinates of P are given by

$$x = VS \text{ and } y = PS.$$

Hence, we have

$$y^2 = 2\,AVx.$$

As in the reasoning the distance between V and A is always the same, AV = k
then $y^2 = 2kx$ and $x = (1/2)k \cdot y^2$.

7. The system of coordinates: from symptoms to equations
A conscious anachronism is introduced to shift from the solid to the algebraic setting, and the standard equation is obtained.

The k constant is introduced by repeating the teacher's words (see "Helping to sum up the process" above)

When k changes, i.e., when the secant plane is translated parallel, the width of the orthotome increases. On the other hand, when k decreases, the width decreases until, when k decreases to zero, we obtain the degenerate orthotome, i.e., the line AK (equation $y = 0$).	8. The meaning of k The meaning of k is constructed by changing it. Also, in the change, the limit case appears and is interpreted correctly.
The obtained symptom characterises whatever point of the section is chosen because, if the distance $AV = k$ is the same and the plane t (where the generator circle lies) is translated parallel to itself, the proof is valid for any other point on the section. Only the length of the arc of the curve has changed.	9. Generalisation to any point of the orthotome The reasoning is generalised to any point of the orthotome by moving one of the fixed planes. Actually, in the movement some lengths are not changed. The final sentence shows the correct interpretation on the model.
Such a section was considered as a solid curve in three dimensions, as it lies on a right-angled cone, and its property is obtained by means of reasoning about two different planes that are perpendicular in space.	10. General comments This final comment shows the detachment of the students from the approach of Greek geometers. The section *was* considered a solid curve. Actually, this term also appears in Descartes' work. This conceptualisation is correctly related by the students to two different issues: the way of obtaining the curve (as a conic section) and the way of proving the symptom by means of figures lying in two perpendicular planes.

3. DISCUSSION

In this section, I shall go through the teaching experiment once more to summarise the relationships between the motives, tasks and operations as they can be detected by looking at the teacher's side, in both the designing and the functioning, and their traces that can be detected by looking at the students' side, in both the oral interaction and the written report. A brief comparison between teacher's side and students' side can be seen in Table 1. This analysis is surely incomplete since I shall refer only to the short excerpts that have been quoted in the text. Some reference to the other available data (Tufo, 1995), however, will be made from time to time.

3.1 The teacher's side

In the section "Motives of the teaching-learning activity for Mathematical Machines" the four motives are listed. The first motive is realised by means of extensive historical introductions, with the reading of original sources, too. The meaning of mathematical concepts cannot be constructed only inside mathematics, so the problem of the *construction of meaning of conics* in the classroom has to be taken as paradigmatic and as representative of different approaches to the problems of space and geometry in the different cultural spaces of different ages (a similar approach can be found in Mancini Proia & Menghini, 1984). The second section of this chapter sketches the kind of historical introductions that are proposed by the teacher. They are not limited to internal history of mathematics but they allude to the wide social and cultural environment where mathematicians of the past lived. Moreover they stress the collective aspects of scientific progress, where to locate individual contributions: hence Euclid, Descartes, Desargues and others are not conceived as isolated geniuses that work in a vacuum, but as exceptional representatives of existing cultural trends.

In this teaching experiment this first motive is especially focused in the initial historical introduction (section "The teacher's lesson: A historical introduction to conics in ancient Greece" and in the presentation of de l'Hospital's work (section "The rest of the experiment").

The second motive contributes to defining the rules of the didactic contract: in some tasks students are allowed, like Greek geometers, to use application of areas and proportions and forbidden to use algebra; in other tasks students are allowed, like post-cartesian geometers, to introduce a system of coordinates on the figure and to write down algebraically the relationships between some line measures; in other tasks only the algebraic equation is considered and so on. The rules are explicitly posed by the teacher at the beginning and recalled during the interaction. One of the effects is that different formats of proof of the "same" statement are considered, so that the meaning of the statement is enriched by the whole activity. The social rules are explicitly related by the teacher to the issues that have been presented in the general historical introductions.

In this teaching experiment, this motive is focused in all the phases of the study of orthotome (sections "Small-group work: The study of the orthotome model", "Students' explanation of the orthotome model" and "The teacher's lesson: Models of the orthotome and the equations of conics"). Some traces are found also in the excerpts we have quoted, at the level of teacher's operations. For instance in the Episodes 1 and 2 of small group interaction, the teacher emphasises the difference between "then" (ancient Greece) and "now" (after Descartes) and clearly states the first task to be solved without coordinates. Later, in the Episode 5, he states again the relationship between ancient and "modern" notation.

The third motive is realised by the sequence of actions in the teaching experiment, with intentional shift to and fro the solid setting and the algebraic setting (and later, with De l'Hospital's work, the mechanical setting too). In the Episodes 1 2 and 5, explicit relationships between the two settings as concerns the description of the curve and the instrument of proving are stated.

The fourth motive is realised by the systematic and intentional recourse to physical models, either static or dynamic ones. Traces of this emphasis are found also in the interaction: in the Episode 2 the teacher explicitly invites students to observe the model and to elicit the hypothetical elements. In the Episode 3, the teacher encourages students to make exploration: he does not use letters for coding points, but points to the model with gestures; he suggest to introduce also not visible elements and to modify in the mind the physical object. In the Episode 5, he shows that the point P can be considered a "generic" point, by introducing movement; he interprets the change of the secant plane both in the model, imagining a movement, and in the equation, imagining a continuous change in the numerical value.

TEACHER'S SIDE			THE STUDENTS' SIDE		
Motives/ Activity	**Actions (examples)**	**Operations (Examples from Small Group Work)**	**Traces in Small Group Work**	**Traces in Final Report**	**Elements of Constructed Meaning**
Relationships between mathematics and other fields of knowledge	Wide historical introductions	Not detailed	—	—	Historical contextuali-sation
Historical contextuali-sation of rules of behaviour	Definition of didactic contract	In Episodes 1, 2 and 5	Not detailed in the excerpts	Sections 1, 2, 3, 7and 10	Of problems concepts and procedures
Multifaceted meaning of conics	Shift to and fro solid, algebraic (and mechanic-al) settings	In Episodes 1, 2 and 5	Not detailed in the excerpts	Sections 7, 8 and 10	Interplay between settings
Dynamic interpretation Principle of continuity	Systematic recourse to physical models	In Episodes 2, 3 and 5	Gesturing Episode 4	Section 8 Section 9	Dynamic interpretation of physical models

Table 1: Comparison between teacher's side and students' side

3.2 The students' side

Now we shall go through the students' protocols to find traces of motives, if any. We shall draw on a very limited set of data, only a couple of excerpts from interaction and the written final report. However, the final report is very long and interesting, because it contains, in a well-ordered style, all the relevant issues of small group interaction. The order of the text is the first relevant feature. This order does not mirror the complex exploration of the model during small group work: having been able to contextualise and to write down the proof on the base of sketchy notes (taken by students standing close to the physical model and not sitting quietly at their desks) proves that the text conveys the constructed meaning.

Traces of the first motive cannot be acknowledged in this limited set of data. Actually changes in the conceptualisation of mathematics are not always explicitly stated by students. They can be revealed by long term listening at their talks; for instance they are evident in the quality of discourse they produce in oral test, in mathematics and in other subjects as well (e.g. history, philosophy, literature and so on).

On the contrary, traces of the second motive are evident. In the Sections 1, 2, 3, 7 and 10 of the written report, explicit reference to history is done again and again. Actually this reference is functionally interlaced with proofs, in the solid setting and in the algebraic setting as well.

Also traces of the third motive are present. In the Sections 7, 8 and 10, conscious anachronism is repeatedly commented: the students are aware that today we can consider the conic section orthotome and the algebraic curve parabola as the same object, but they are conscious that both the definition and the instruments of proof are quite different and historically contextualised.

The fourth motive is revealed by different sets of data: the large amount of gesturing in small group interaction (some examples are in the Episode 4 of small group interaction); the conscious recourse to movement in interpreting the meaning of k (Section 8 of the written report) up to the analysis of the limit case ($k = 0$) that had not been considered during small group interaction; and the conscious recourse to movement in extending the property to any point of the curve (Section 9 of the written report). Actually some of these points had been hinted at by the teacher very quickly: the independent written reconstruction is so clear and neat that the students are supposed to have internalised joint activity with the teacher.

If this analysis is correct, the meaning of conics that is now constructed by students is more complex than in standard classrooms. At least three elements that are usually lacking enter as constitutive parts of the meaning:

1. the historical contextualisation of problems and concepts,
2. the relationships between the solid setting and the algebraic setting (and, in the last step, the mechanical setting),
3. the dynamic interpretation of physical models to guess conjectures and to guide the construction of proofs

The development of this complex meaning draws on two special choices that have been made in designing the teaching experiments: the recourse to historical sources and the activity on physical models.

As concerns the former, it is clear from the text that, at the beginning, the students are (consciously) in the solid setting, while later they shift (consciously) to the algebraic setting. The interplay between the two settings is evident when they interpret the change of the parameter k (up to the limit case $k = 0$) as a parallel translation of the plane. Hence the model is studied with the introduction of a conscious anachronism, that shows limits and advantages of each approach. In the process of solution of the given problem, there is a shift from one setting to another and from one time period to another; there is a continuous change of the objects, that are identified by means of a process of regressive appropriation: what they know now on conics allow them to go to and fro the individual sections and the individual settings, using the most advantageous tools for proving. In this process the historical reconstruction of meaning is not a way to motivate students or to embellish the problem but is a fundamental object of the teaching learning activity.

The presence of the physical model is essential in the process of solution. The observation of the small group work (during which the proof has been built) has shown a large volume of visual tactile activity (e.g., gesturing, pointing at the model) while the discussion was going on. This is an invariant aspect of all the laboratory sessions. Yet a transformation of the physical object into an ideal object is observable, as the study of the physical object is done with reference to Euclidean theory of proportions and to analytic geometry. This process between the physical object and the ideal object is dialectical: at the end the interpretation of the values of the parameter k is done on the physical object; besides the generalisation of the properties to a whichever point of the section is done again on the physical object. The translation of the planes (the plane of the section for k, and the base plane for the generalisation) is done looking at the physical object (where the plane are fixed) and moving the hands up and down (a further analysis of a similar process on a linkage is done in Bartolini Bussi, 1993).

3.3 Open problems

In this paper we have introduced some elements of an historical analysis of conics (to be meant as a paradigmatic example of geometrical concepts) to claim that their present meaning as objects of the knowledge to be taught is not one-sided but ground in different settings determined by the processes of studying conics in different time periods.

The didactic problem is how to introduce in the classroom the epistemological complexity suggested by this historical study. A pragmatic solution is offered by the research project *Mathematical Machines*. But this solution opens two different kinds of problems, concerning:

1. the microstructure of teaching-learning activity in the classroom,
2. the relationships between the elements of the context (mainly historical sources and physical models) and the student processes.

The former problem has to be meant as the tentative modelisation of the teaching-learning activity at the level of actions-operations (Leont'ev, 1978). As we adopt a Vygotskian perspective on the teaching learning process, we claim the need of considering phases of joint activity between the teacher and the students (Bartolini

Bussi, 1993); however the complexity of the processes does not allow us to make a priori analysis of them.

The latter problem concerns the two main distinctive characters of the context, that makes it different from the ordinary one, namely the systematic introduction of physical models and of historical sources. The data we have collected until now show the extraordinary effect they can have on students' construction of meaning. We are interested in detecting more precisely the roles and the conditions of functioning of both in this process. Physical models emphasise the role of visual tactile activity in a way that appears quite different from software tools: the stiffness of physical models forces students to make experiments in the mind and to anticipate results that cannot be controlled empirically, while the flexibility of dynamic software such as Cabri rather invites students to make concrete experiments and to observe their effects. Historical sources fosters student self location in the collective cultural activity of mankind. Both aspects are studied in the activity theoretical approach drawn on the work of Vygotskij, Leont'ev and others (see for instance Tikhomirov, 1984 for the former; Otte & Seeger, 1994 for the latter). So, the main aim of our research group now is to look for a comprehensive theoretical framework that allows us to interpret the relationships between these two characters and student construction of meaning and to design further teaching experiment where to realise and analyse such relationships (Bartolini Bussi et al., forthcoming).

ACKNOWLEDGEMENTS

The machines have been built and used in the classrooms by Annalisa Martinez, Marcello Pergola and Carla Zanoli. I wish to thank them for their invaluable more than decennial cooperation and for the permission of including in this paper parts of unpublished documents and worksheets produced for internal use of the research group. The laboratory of Mathematical Machines has recently moved to the Department of Mathematics of the University of Modena and Reggio Emilia (http://www.mmlab.unimore.it). Activity is now funded by the University and the MIUR (Italian Ministry of Education, University and Research, Grant nr. 2003011072).

REFERENCES

Bartolini Bussi, M. G. (1993). Geometrical proof and mathematical machines: An exploratory study. In I. Hirabayashi, N. Nohda, K. Shigematsu & F. L. Lin (Eds.), *Proceedings of the 17th International Conference for the Psychology of Mathematics Education* (vol 2, pp. 97–104). Tsukuba (Japan): PME Program Committee.

Bartolini Bussi, M. G. (1998). Drawing instruments: Theories and practices from history to didactics. *Documenta Mathematica – Extra Volume ICM 1998, 3*, 735–746.

Bartolini Bussi, M. G. (2000). Ancient instruments in the mathematics classroom. In J. Fauvel & J. van Maanen. (Eds), *History in mathematics education: The ICMI Study* (pp. 343–351). Dordrecht: Kluwer.

Bartolini Bussi, M. G. (2001), The geometry of drawing instruments: Arguments for a didactical use of real and virtual copies. *Cubo Matematica Educacional, 3*(2), 27–54.

Bartolini Bussi, M. G. & Mariotti, M. A. (1999). Instruments for perspective drawing: Historic, epistemological and didactic issues. In G. Goldschmidt, W. Porter & M. Ozkar (Eds.), *Proceedings of the 4th International Design Thinking Research Symposium on Design Representation* (Vol III, pp. 175–185). Cambridge: Massachusetts Institute of Technology & Technion – Israel Institute of Technology.

Bartolini Bussi M. G., Mariotti M. A. & Ferri F. (Forthcoming). Semiotic mediation in primary school: Dürer's glass. In H. Hoffmann, J. Lenhard & Seeger F. (Eds.), *Activity and sign: Grounding mathematics education*. Dordrecht: Kluwer.

Bartolini Bussi, M. G., Nasi, D., Martinez, A., Pergola M., Zanoli C., Turrini M., Vacca M., Malagoli S., Ricchetti A., Dinelli P. & Greco, S. (1999). *Laboratorio di Matematica: Theatrum Machinarum*, 1° CD rom del Museo. Modena (Italy): Museo Universitario di Storia Naturale e della Strumentazione Scientifica http://www.museo.unimo.it/theatrum.

Bartolini Bussi, M. G. & Pergola, M. (1994). Mathematical machines in the classroom: The history of conic sections. In N. Malara & L. Rico (Eds.), *Proceedings of the First Italian-Spanish Research Symposium in Mathematics Education* (pp. 233–240). Modena (Italy): Dipartimento di Matematica, Università di Modena.

Bartolini Bussi, M. G. & Pergola, M. (1996). History in the mathematics classroom: Linkages and kinematic geometry. In H. N. Jahnke, N. Knoche & M. Otte M. (Eds.), *Geschichte der Mathematik in der Lehre* (pp. 39–67). Goettingen: Vandenhoeck & Ruprecht.

Bos, H. J. M. (1981). On the representation of curves in Descartes' *Géométrie*. Archive for History of Exact Sciences, *24*, 295–338.

Bourbaki, N. (1974). Eléments d'Histoire des Mathématiques. Paris: Hermann.

Coolidge, J. L. (1945). A history of the conic sections and quadric surfaces. Oxford: Clarendon Press.

Halby, M. (1988). Drawing instruments 1580–1980. London: Sotheby's Publications.

Hanna, G. & Jahnke, H. N. (1993). Proof and applicaton. *Educational Studies in Mathematics, 24*(4), 421–438.

Leont'ev, A. N. (1978). *Activity, consciouness, personality*. Englewood Cliffs (USA): Prentice Hall. (Original edition 1975.)

Mancini Proia, L. & Menghini M. (1984). Conic sections in the sky and on the earth. *Educational Studies in Mathematics, 15*, 191–210.

Otte, M. (1993). Towards a social theory of mathematical knowledge. In C. Keitel & K. Ruthven (Eds.), *Learning from computers: Mathematics education and technology* (pp. 280–306). Berlin: Springer.

Otte, M. & Seeger, F. (1994). The human subject in mathematics education and in the history of mathematics. In R. Biehler, R. W. Scholz, R. Straesser & B. Winkelmann (Eds.), *Didactics of mathematics as a scientific discipline* (pp. 351–365) Dordrecht: Kluwer.

Pergola, M. & Zanoli C. (1994). 'Macchine matematiche': Note su un progetto di ricerca didattica. *L'insegnamento della matematica e delle scienze integrate, 17*, 733–751.

Pergola, M. & Zanoli, C. (1995). Trasformazio ni geometriche e macchine matematiche. *L'insegnamento della matematica e delle scienze integrate, 18*, 689–714.

Tikhomirov, O. (1984). *The psychology of thinking*. Moscow: Progress Publishers.

Tufo, A. (1995). *Storia e modelli fisici nella didattica delle coniche: il caso dell'orthotome* Modena (Italy), Università degli Studi di Modena, Final thesis.

ROLF BIEHLER

RECONSTRUCTION OF MEANING AS A DIDACTICAL TASK: THE CONCEPT OF FUNCTION AS AN EXAMPLE

The meaning of a mathematical concept differs in different contexts. We can find different practices related to the same mathematical concept, such as the concept of function. Physicists have a practice different from that of mathematicians. Qualitative functional thinking is necessary in many vocations, however, this is a fairly different practice from that of academic sciences with their explicit use of symbolic mathematical notation. Commonly, uses will differ even if people use the same definition of a concept. However, spheres of practice may also differ with regard to definitions of a mathematical concept. The notion of a functional relationship between magnitudes may still be much used in physics, whereas mathematicians tend to use a more general notion of correspondence between sets. If we speak of different meanings of "the same" concept, we can further analyse differences and commonalities. How can didactics of mathematics cope with these many meaning differences? The mathematics classroom should not be a closed and self-reproducing system developing its own concept meanings. Rather, the meanings that are to be constituted in the classroom should be related to practices and meanings outside school. But what are the important points of orientation?

All the various spheres of practice (academic mathematics is one of them) in which mathematics is used are, in principle, relevant sources of meaning for general education. What dimensions of meaning of a concept should curriculum designers ideally take into account? The meaning and the importance of the concept within the theoretical network of academic mathematics, its historical genesis and development, its uses for problem solving inside and outside mathematics, its prototypical interpretations, its roots in everyday thinking and language as well as different tools and representations for working with the concept are relevant. How these sources are exploited and given relative weight to is dependent on the social meaning attributed to mathematics education. The social meaning varies. For instance, the traditional German *Gymnasium* had to prepare students for university studies, and the traditional German *Volksschule* had to prepare students for various vocations (artisans, workers etc.). The meanings of mathematical concepts that were selected for the various student groups differed very much according to the various social functions of schools and according to assumptions concerning what these students were able and willing to learn under given societal and schooling conditions.

We know that teachers are important agents in the classroom constitution of mathematical meaning. The implementation of curricula by teachers is shaped by their beliefs, in particular, by what they consider to be important aspects of a concept's meaning. If teachers themselves share some of the rich meanings which are implicit in curriculum material and serve as a background for its design, it is more probable that the intended meanings of the concepts will be implemented in the classroom. If teachers are not explicitly trained or educated in this respect, they may tend to base their teaching on the meanings they have acquired elsewhere, namely the traditional meanings of school mathematics, or on the meanings of concepts they have acquired during their academic studies in mathematics—if they have had an academic mathematics education and still consider this orientation the most important source for their teaching. When students study academic mathematics, they are confronted with meanings of concepts that can be considered only as part of the overall meaning landscape. The many uses of the concepts in various disciplines and in societal practices (and also in history) can be considered as part of a very comprehensive meaning landscape of that concept. But usually, these uses and practices are not part of the consciousness of mathematics students, professors and educators.

As mathematics education, however, has to base its curricular decisions on a broader picture of mathematics than that of academic mathematics, we consider the *reconstruction of meaning,* the development of a *synthesising meaning landscape* of a mathematical concept to be an important task for the didactics of mathematics that could serve as a theoretical background for curriculum design and implementation. We will also speak of a *didactically reconstructed intended mathematics for schools.* In this paper, we will argue in favour of a more systematic approach to this problem, illustrating and exemplifying our own ideas with regard to the concept of function. In some points in educational history, we can well identify interesting attempts to construct intended mathematics for schools as a referent for constituting the *knowledge to be taught* in Chevallard's (1985) sense. We will start with discussing some attempts below that will also show that it is usually not just "academic mathematics" that functions as a referent for constituting knowledge to be taught.

1. MEANING OF FUNCTIONS IN THE CONTEXT OF DIDACTICALLY RECONSTRUCTED MATHEMATICS

1.1 Examples of didactically reconstructed mathematics

Reconstructions of meanings of functions were often embedded in more general attempts to reconstruct the meaning of mathematics in the context of reform attempts in mathematics education.

Well-known historical examples for such a reconstruction are Felix Klein's books on "Elementary mathematics from a higher standpoint" (Klein, 1925a; Klein,

1925b; Klein, 1933) where he described and synthesised a view and selected a content of mathematics for German *Gymnasium* teachers, who had already good knowledge in mathematics. A value system is implicit in his books. It is related to the reform efforts of restructuring school in the direction of giving more emphasis to geometrical aspects of meaning (intuition, *Anschauung*) and to applications. A reintroduction of geometrical and visual aspects was regarded as necessary for school mathematics and for users of mathematics at the same time. The arithmetisation and formalisation that had led to banning geometry from the foundations of mathematics was not considered to provide an acceptable basis. A particular expression of this reform was the emphasis on "functional thinking" as one of the major goals of school mathematics. Functional thinking was considered a fundamental idea that should integrate pure and applied aspects of mathematics and legitimise the introduction of calculus into the senior secondary curriculum. Calculus was considered the top level of functional thinking that should be taught in the senior grades of secondary schools but that had to be adequately prepared in junior grades prior to that. Klein' s books are a good prototype of reconstructed mathematics because they not only develop a "philosophy of mathematics", but rather a view of mathematical content from a certain "philosophical" perspective that is more or less explicit. Klein introduced some epistemological distinctions, namely the distinction between "precision mathematics" and "approximation mathematics" as a way to describe the difference between the ideal and exact world of mathematics and mathematics applied to reality and to *Anschauung*. Typically, Klein does not just present "school mathematics" but goes far beyond this level with regard to the content treated.

Another big historical event for the function concept in mathematics education was the new math reform where functions were reconstructed as examples of the general concept of mapping, or as a specific relation. New meanings were derived from this embedding, whereas traditional aspects of meaning as "relations between magnitudes" were devalued. We can interpret the dozens of books on "new math for teachers and parents" as attempts to constitute a type of didactically reconstructed mathematics, although the writers would have thought of it just as of an elementarised academic mathematics. The latter illusion is quite understandable. If we only look at their concept definitions and theorems, then their mathematics will often appear only as a subset of academic mathematics. But if we include looking at domains of application, at the surrounding conceptual structure of a concept, and at the tools and means of representation used with a concept, we begin to see the differences.

A recent example of what I would consider a type of didactically reconstructed mathematics is the intended school mathematics constructed for the NCTM Standards (National Council of Teachers of Mathematics, 1989). This book, however, is already very much concerned with intended school teaching and learning processes. Maybe we can consider the book edited by Steen (1990) a description of the related didactically reconstructed mathematics as such, and as somewhat more separated from teaching and learning methods. The new NCTM's didactically reconstructed mathematics and other contemporary ones often make connections to the new humanistic and descriptive views of mathematics that include the "social dimension", problem-solving and the investigative nature of mathematics. (Davis & Hersh, 1980; Ernest, 1994). It is not clear how far mathematicians will see this characterisation of mathematics as an extension of their own view, or whether we are faced with an example of an artificial mathematical culture whose relation to academic mathematics is still pretty opaque.

In the context of this reform movement, reconstructions of the meaning of the function concept have been performed (Romberg, Fennema & Carpenter, 1993). A didactical analysis concerning the meaning of the concept of function in various mathematical practices or cultures is however still lacking. This deficiency is pointed out by Williams (1993, p. 315) in relation to the above book: "What we have instead is a description of a unique ethereal culture that, it can be argued, does not currently exist. It is well described by the *Curriculum and Evaluation Standards for School Mathematics* of the National Council of Teachers of Mathematics" (see also Biehler, 1994).

Attempts at reconstructing meanings of the function concept with regard to school mathematics after new maths are prevalent in other countries, too. A prominent example of the search for meaning is Freudenthal's (1983) *Didactical Phenomenology of Mathematical Structures*. He states as goals of his program:

> *Phenomenology* of a mathematical concept, structure or idea means describing it in relation to the phenomena for which it has been created, and to which it has been extended in the learning process of mankind, and, as far as this description is concerned with the learning process of the young generation, it is *didactical phenomenology*, a way to show the teacher the places where the learners might step into the learning process of mankind. (p. ix)

From this general approach, he develops a didactical phenomenology of functions (pp. 491–578). These reconstructions are related to reform attempts in the Netherlands under the conception of *realistic mathematics education.*

1.2 Research for meaning reconstruction and the complementarity of the function concept

There have been several studies concerning the meaning of the function concept that are not so closely related to reform movements. Vollrath's paper (1989) provides an example of synthesising aspects of the meaning of functions that were particularly discussed in Germany. Sierpinska's (1992) study on the meaning of functions can also be considered as intending a re-construction. She states the objectives of such research:

> In our attempt to define the basic conditions for understanding functions we shall be guided by an exploration of the reference of the definition of this notion. We shall ask ourselves what is this reality this definition refers to, what objects are there to be identified, discriminated between, what kind of orders can be found that would bring about the enlargement of reality by way of insightful generalisations and syntheses. (p. 30).

The analysis of regularities in relationships between changing magnitudes constitutes an important source of functions, i.e. a central part of its meaning. This is substantiated by Sierpinska's (1992) contribution analysing the concept's historical development in mathematics, which was related to uses in physics and geometry who were major "partners" in development. One of her results is the suggestion that "students must become interested in variability and search for regularities before examples of well-behaved mathematical elementary functions and definitions are introduced." (p. 32). This attitude may constitute an epistemological obstacle for teachers who have been "brought up" in a "pure-mathematics-culture". Sfard (1992) emphasises the dual nature of mathematical concepts as process and object and develops the thesis of the primacy of the operational origin of mathematical concepts. Structural notions emerge by reification later. The computational process of starting from a number x and calculating a resulting value y is, according to Sfard, the major source of the function concept.

These two positions really point to two sources of the notion of function admirably expressed by the mathematician Hermann Weyl (my translation, R.B.):

> Historically, the concept of function has a double root. Leading up to it are firstly the 'naturally given dependencies' ruling the material world which consist, on the one hand, in the fact that states and constitutions of real things are changeable in time, and on the other in the causal connection between cause and effect. A second root quite independent of the first lies in the arithmetico-algebraic operations. According to this, the analysis of old had in mind an expression which is formed from the independent variable by applying the four species and some less elementary transcedents a finite number of times, though these elementary operations have never been clearly and completely designated and historical growth has always pushed beyond to closely set boundaries without the agents of this development realising this every time. The point where these two sources which are at the outset quite foreign to one another begin to relate is the concept of the natural law. Its essence consists in the very fact that the natural law represents a naturally given dependency as a function constructed in a purely conceptual-arithmetical way. Galileo's laws of falling bodies are the first important examples. The modern growth of mathematics has led to the insight that the special algebraic principles of construction on which the analysis of old was based are much too narrow for a logico-natural and general development of analysis as well as when the role is considered which the function concept has to assume for the recognition of the laws governing what happens in the field of matter. General logical principles of construction must replace those algebraic ones. (Quoted in Weyl, 1917, p. 35-36)

I have taken the quote from the book of the IDM-Arbeitsgruppe Mathematiklehrerbildung (1981), who used it as an illustration for what they call the complementarity of the concept of function. The function concept is an excellent example of the complementarity of concepts in mathematics (Otte, 1984). In recent years, the didactical analysis of the concept of function has led to a revival of various characterisations of complementary aspects of functions. An early reference to the complementary aspects of the function concept as a mathematical object and as a thinking tool is Otte and Steinbring (1977). Other complementary characterisations are descriptive-relational vs. algorithmic-constructive (Richenhagen, 1990), geometrical-set theoretic-extensional vs. algebraic-analytical-intensional (Steiner, 1969), process and object respectively dynamical mapping vs. static relation (Sfard, 1992), co-variational versus correspondence aspects (Confrey & Smith, 1994), and, similarly, Vollrath (1989) who emphasises the distinction between horizontal (correspondence) and vertical (co-variation) aspects of functions. Concepts of complementarity have proved useful for empirical and constructive research on functions (Dubinsky & Harel, 1992; Romberg, Fennema & Carpenter, 1993).

2. RECONSTRUCTION OF MEANINGS OF THE CONCEPT OF FUNCTION

2.1 The context: Developing teachers' knowledge

I will now go into more detail concerning the function concept. As usual, ideas are shaped by the context they were developed in. The following ideas were developed in connection with pre-service courses for teachers on "the concept of function and functional thinking". The teacher-students already had a good mathematical background, but the intention was to enable them to reflect, enrich, and restructure the meaning they associated with the concept of function with regard to mathematics education. My selection of aspects was to emphasise a teaching of functions with technological support, applications outside mathematics and the general idea of functional relationship as contrasted to the limited view of functions normally taught in school. In particular, I will point to the meaning differences between functions in academic mathematics and what I consider as important for school teachers.

Empirical studies concerning teachers' knowledge of functions have to be based on an overall conception of knowledge on functions. For instance, Ruhama Even' s (1989, 1990, 1993) empirical study on teachers' knowledge of functions is based on an integrative analysis of what is considered as the meaning of the function concept in some part of the relevant didactical literature:

> As a result of this integration, six aspects seemed to be critical components of subject matter knowledge required to teach functions:
>
> - What is a function? (including image and definition of the concept of function, univalent property of functions, and arbitrariness of functions).
> - Different representations of functions.
> - Inverse function and composition of functions.
> - Knowledge about functions of the high school curriculum.
> - Different ways of approaching functions: point-wise, interval-wise, globally, and as entities.
> - Different kinds of knowledge and understanding of functions and mathematics. (Even, 1989, p. 212)

Many of Even's detailed results are interesting and point to a need to change teacher education in this area. However, her study represents a view of functions from a certain didactical discussion, whose strengths and weaknesses carry over to her study.

We have to broaden the perspective. I will describe a more extended meaning landscape for mathematics educators. It should also serve as a basis for further discussion on which aspects of the landscape are most important for teachers.

I will briefly discuss exponential functions as an example. Figure 1 contains a sketch of a semantic landscape for exponential functions. This is an example that has been considered in other recent projects, too (Confrey, 1991; Confrey & Smith, 1994). The exponential function is to provide a concrete example for important relations in the landscape. The network should give an impression of the conceptual complexity. I will not explain all individual elements and their importance in detail. That would be beyond the scope of this paper. Some comments must suffice. The picture contains theoretical mathematical aspects (difference, differential and functional equation, isomorphism between addition and multiplication, power series and number systems), relations to the dynamical systems and growth and decay processes, relations to other growth functions, to discrete models (geometric series), computational aspects (tables → slide rules → algorithms), relations to statistics and data analysis (curve fitting, log scales, data graphs), domains of application (radioactive decay and population explosion) and related general concepts relevant in applications (prediction, explanation, and model).
As compared to the normative view concerning teachers' knowledge on exponential functions on which Even (1989) based her study, the content of the above semantic network appears to be very ambitious as content for teachers to be learned. Compared with what secondary teachers have to learn in mathematics, far away from the elementary level, the landscape seems to be quite acceptable. Present teacher education does not yet provide sufficient preparation to enable teachers to develop such a complex system of meaning for themselves—in the first place. Even if all individual elements of the landscape were present in the teacher's mind, it is questionable whether he or she sees it as a whole, as a highly interrelated network of meaning as a background knowledge and meta-knowledge as a basis for teaching exponential functions in school.

As mathematical teacher education must capitalise on the teachers' ability to extend and reorganise their professional knowledge during their future life, we have to think of adequate measures to ensure that teachers not only improve their practical knowledge of teaching methods by way of experience, but also actively extend their mathematical meanings beyond those they have already learned during their studies within academic mathematics.

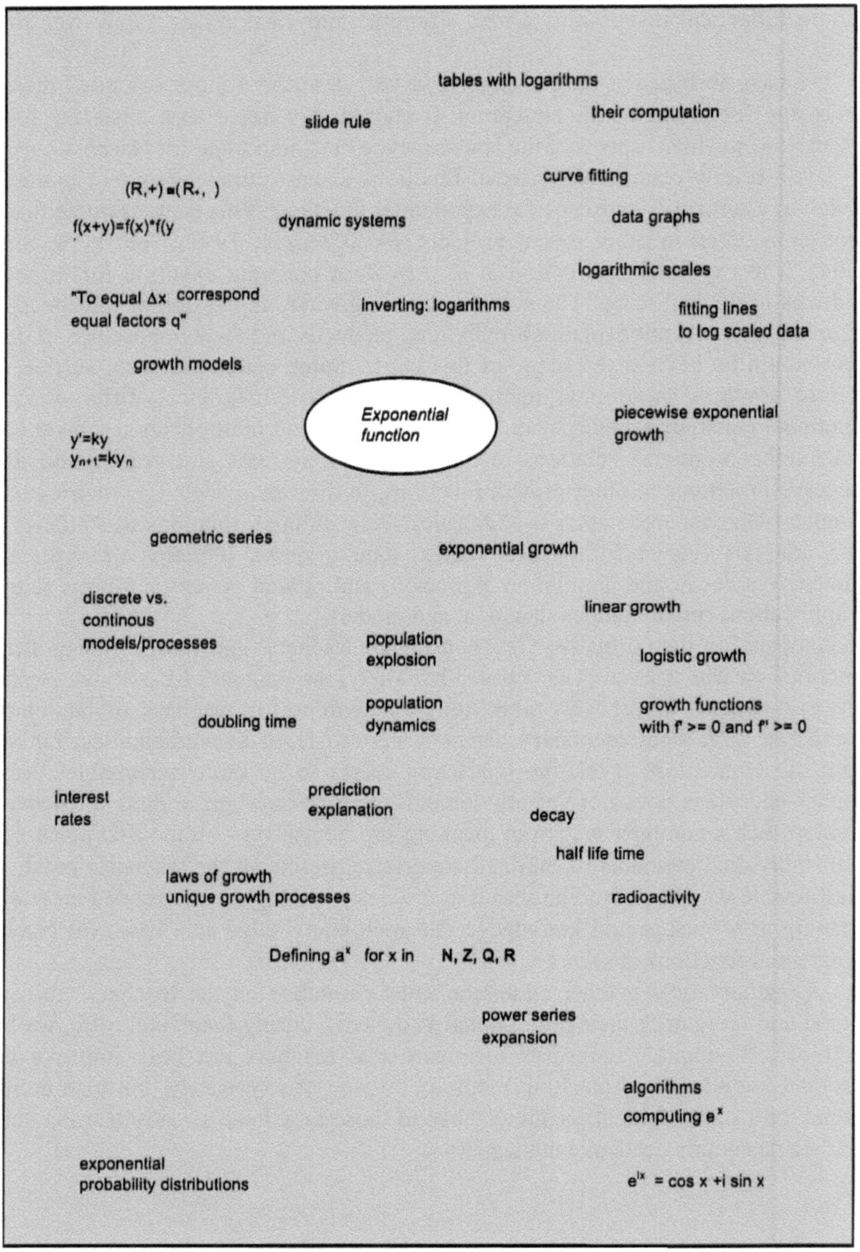

Figure 1: Semantic landscape for exponential functions

2.2 The meaning landscape in general

We need some distinctions concerning the conception of "meaning" that will enable us to structure our meaning landscape. I will use a variant of the epistemological triangle, which Steinbring (1994) uses to discuss the meaning of mathematical concepts. In Figure 2, I have drawn a variant that is most adequate for my current purpose. The epistemological triangle is based on the belief that the domains of application (a concept's uses inside and outside mathematics) are constitutive for what we may call meaning of a concept. Also the relation to other concepts, its role within a conceptual structure (a theory) and the tools and representations available for working with a concept are constitutive parts of the meaning. These dimensions constrain the problems for which the concept can be used. The epistemological triangle interpreted that way also implies a time dependence of meaning. Meaning may change by new applications, by new conceptual relations, or by new representations. I consider conceptualisations of knowledge like the *conceptual fields* (Vergnaud, 1990) and the *semantic fields* (Boero, 1992) as conceptualisations that are similar (see "Meanings of Meaning of Mathematics" in this volume for a more detailed account). Table 1 is an attempt to structure elements of a network according to the meaning components of the epistemological triangle: relations to other concepts (inside and outside mathematics), representations, and applications. I have used Sierpinska's (1992) study and Freudenthal's (1983, pp. 491–578) phenomenology as one of my sources for developing this "semantic landscape" for functions. I have added aspects that come from the practice of using functions in connection with statistics (in italics) and related to the new technologies (underlined), which are not covered by Freudenthal's and Sierpinska's analyses.

The discussion of all the elements of the table, the relation between elements and what teacher-students do know or should know about that and why cannot be done within the scope of this paper. In the following sections, I will only comment on some aspects.

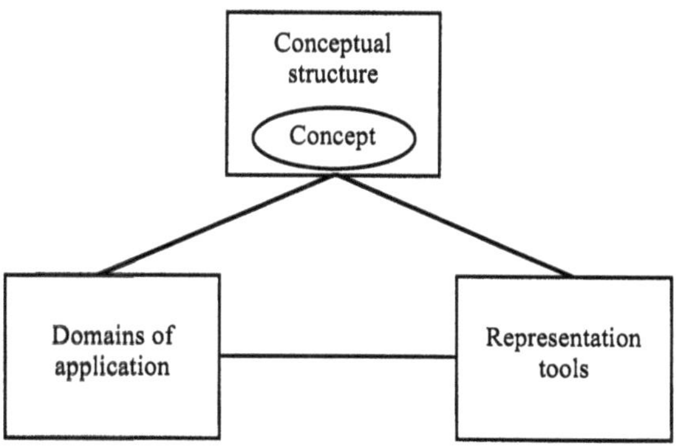

Figure 2: A variant of the epistemological triangle

2.3 Relations within mathematics

A didactically reconstructed meaning landscape has to overcome a compartmentalisation that is typical for the experience of a mathematics teacher-student. Algebra, calculus, differential equations and statistics are different courses in the university life of a mathematics student. Moreover, an average stochastics course in Germany would probably not cover regression and correlation. It is even more likely that students are not aware of the fact that the notions of correlation and conditional expectation can be regarded as generalisations of the concept of function, as a tool for analysing relations between magnitudes or "variables", as a statistician would say. Another generalisation of the operational aspect is that functions can be defined by algorithms or computer programs—extending the repertoire of algebra and of analytical expressions.

Generally, differential equations belong to a different cognitive compartment than functions, and students are not aware of the intimate historical relations existing between the emergence of the function concept and differential equations. The idea of an "unknown" function that is characterised by equations was pretty important for the constitution of functions as mathematical objects of study. Moreover, there are relations relevant to school mathematics that are no longer paid attention to in academic mathematics, in which a certain mathematical practice is already assumed. The different uses and meanings of variables are a good example: their use as unknowns in the context of solving equations, their use in describing rules for functions, and their use as symbols that signify variable magnitudes.

2.4 Representations

Computers provide plentiful new representations for functions that can be valuable for meaning development and for extending the range of applications. A reflection about the scope of different representations is something that has to be stimulated in teacher education courses. Often however, teachers have not yet become part of a practical mathematics culture where computer use for problem solving (similar to the practice of engineers) is common. This is why reflection and new experiences are necessary. I will discuss some aspects in more detail.

2.4.1 Language of functions and graphs

An important didactical idea for developing the function concept beyond algebraically defined functions consists in asking students to qualitatively sketch curves that describe features of processes (see Höfler, 1910, for a such an approach in history; and Swan, 1982, for a modern conception). Höfler was part of Klein's reform movement that intended to put relatively more emphasis again on the geometrical aspect of the complementary duality of functions. If students are asked by teachers to sketch a curve of the dependence of the water level upon time when various bottles are uniformly filled, then teachers should also know how this problem could be solved with advanced mathematical means: The sectional area as a function of time or height has to be integrated to get the volume as a function of time, and so forth. If spherical bottles are being used, it should be clear how to determine the volume of parts of a sphere. Integration in this context is a special case of solving a differential equation and teachers should be aware of the relation of these elementary integration tasks to differential or difference equations.

Related concepts in mathematics	**Related concepts in science and applications**
(mathematical) relation	law
univalence of relation	causal relation
asymmetry of variables	dependence
variables (unknowns) in algebra	interdependence, interaction
equation	[← curves]
proportionality	[← motion/change in time], change in general
algorithm	variable magnitudes
differential equation	relations between magnitudes
functional equation	equation between magnitudes
sequence	data tables
mapping, operator	*data graphs*
correlation	*time series*
conditional distribution	*strength of a relationship*
regression	machine
	constructed relation

Representations	**Applications**
symbolic:	prediction
algebraic equation	description
analytical expression	interpolation
implicit definition, "properties"	extrapolation
algorithm	*data reduction*
computer program	determining (*estimating*)parameters
manipulable object in software	interpreting parameters
graphs:	modelling
standard Cartesian graph	range of validity
computer based Cartesian graphs	*univalence as an idealisation*
(manipulable scales and zooming)	*deviation from model*
various other graphs	*goodness of fit*
tables:	dynamical systems
standard tables	
interactive spreadsheet tables	
multiple linked representations	

Table 1: Elements of the semantic landscape of the function concept

The idea of a function as an arbitrary free hand curve underlies this didactical approach. It was a central idea in history besides considering functions given by expressions. The arbitrariness of the free hand curve is more limited than the arbitrariness of the so-called Dirichlet definition of an arbitrary correspondence. Klein (1933) intended to mathematise the intuitive notion of "free hand curve" from a mathematical point of view. Klein obviously felt that this notion may provide a more adequate background for functions in school mathematics than the more general definition of Dirichlet. This is an interesting aspect of Klein's didactically reconstructed mathematics, which, by the way, did not really survive in history.

Students may ask their teacher whether it is possible to find a "formula" for every free hand curve. Teachers should know something about the problem of finding analytical expressions for arbitrary (continuous) curves, i.e., that the concept of algebraic formula had to be extended in history in the direction of "analytical expressions", which included infinite series, integrals and other things. In the sense of the earlier quotation from Hermann Weyl, teachers may begin to appreciate that the modern mathematical language, that algorithmic and programming language representations of functions have again extended the repertoire of constructive building blocks usable to reproduce the various relations that can be found in the real world.

The representation of a curve by a formula is also a relevant question when the shape of all kinds of things is to be mathematically expressed: the field of computer graphics provides myriads of applications for this basic idea. In summary, teachers have to be enabled to add and integrate meaning to functions from their knowledge of several separate courses of academic mathematics they may have attended.

2.4.2 Different representations of functions

Working with different representations and relating them to each other is regarded as a basic element of a meaningful teaching and learning of functions. A "classical" aspect is the geometrical meaning of the coefficients of standard functions such as parabolas. For instance, Even (1989, pp. 127) assesses teacher-students' knowledge in this domain. However, interpreting the "subject matter meaning" of coefficients would be a further, often neglected step. Also, more complicated functions are also relevant. For instance, the following equation for logistic growth has to be interpreted according to the various coefficients (*K*, for instance, is the level of saturation).

$$f(x) = \frac{K}{1 + e^{-b(x-a)}}$$

This family of functions can be parameterised quite differently, and teacher-students have only limited experience in choosing an algebraic representation so as to make it better interpretable. Also, how a function as a whole "depends" on its parameters is an important dimension of meaning that is needed in various domains of application.

2.4.3 Functions and equations

The following equations express typical relations between quantities in geometry.

$$A = a \cdot b$$

$$C = 2\pi r$$

Equations in other domains of application are similar. An example is the basic equation in electricity between intensity of current, resistance and voltage:

$$V = R \cdot I$$

An important element of practice consists in interpreting these equations as relations between quantities without any unidirectionality of the function concept in the first place. Their interpretation as functions, however, is also very important but it is not unique. Each equation can be interpreted in various ways. The electricity formula can be interpreted, among other things, as

$(I,R) \mapsto V$, where $V = I \cdot R$, as a function of two variables
$I \mapsto V$, where $R = const.$,
$V \mapsto I$, where $R = const.$,
$R \mapsto I$, where $V = const.$, and $I = \frac{V}{R}$
$R \mapsto V$, where $I = const.$, and $V = I \cdot R$

Each interpretation may correspond to a different situation or problem in reality. Similar interpretations can be done with the geometrical formulas. There are several studies showing that such a flexible functional interpretation of formulas is an important prop required to understand the scientific use and meaning of formulas (for instance, Kriesi, 1981). This qualification is also relevant in pure mathematics where it may pay to see a formula from a new functional perspective. Re-evaluating and re-discovering this practice for school mathematics was also an achievement of didactical research (see Harten et al., 1986).

2.4.4 New tools for working with functions

Software broadens the range of operations that can effectively be performed with functions. Geometrical aspects of meaning conquer more importance. Some of the necessary shifts and problems in teachers' knowledge in these new conditions have been studied by Zbiek (1992).

Teachers have to be also aware of the following problems. Using software for dealing with mathematical notions and theories leads to the problem of a "computational transposition" (Balacheff, 1993): there are shifts of meaning due to transforming knowledge to another representational system. Software, for instance, usually does not handle functions as Platonic objects. They could be represented as finite list of numbers or pixels that approximate the exact values. A generating algorithm could lie behind it, or not. In addition, every software tool has its own set of admissible operations with functions that also determine the "meaning" of functions in this context. The computational transposition can be the source of "meaning conflicts" when students are working with the software. Winkelmann

(1988) provides an instructive overview about the many different implementations of functions in various pieces of software.

2.5 Central historical domains of application

Motion and curves formed essentially important domains of application in history. In the above Table 1, the place of curves and motion could be within mathematics or within applications, just as kinematics and geometry have the same status as applied mathematics from the point of view of formal symbolic mathematics. This leads to the general question what kind of historical knowledge on the development of meaning of a concept including epistemological obstacles is helpful and necessary for the didactics of mathematics and for teachers. Various approaches to this problem can be found elsewhere (Jahnke, Knoche, & Otte, 1996); a particular use of a historical context for meaning development is made by Bartolini Bussi (this volume). Historical domains of application may contribute to making the state of current academic mathematics more understandable than contemporary concept applications do, which already depend on that level of development. Teachers' knowledge on historical domains of application may have a specific cultural value as such and contribute to guaranteeing a cultural continuity in meaning transmission.

In the literature known to the author, there seems to be a certain bias in the historiography of the function concept, namely concentrating on the "pre-history" that led to the modern Dirichlet or Peano (set theoretic) definition of functions. From the standpoint of applied mathematics, other definitions and meanings were still co-existent. Also, the relation to related concepts such as correlative relation as contrasted to functional relation seems to be usually neglected in the historiography of the function concept.

2.5.1 Functions and curves

Curves were one of the key contexts in which the concept of function emerged. The univalence requirement and other factors like the relative marginal role of curves in new math as compared to other instances of the general concept of "mapping", led to a situation where curves and functions became quite separate things in mathematics education. Cartesian function graphs are, now, just one representation of the concept of function, whereas the idea that the concept of function is used to study curves, which are genuine geometrical objects with an existence of their own, independent of the concept of function, was nearly forgotten or at least devalued in mathematics education. The forgotten meanings and relations had to be reconstructed in didactical research (see Weth, 1993). Computers contribute to the possibility of using kinematic curves presented by animated computer graphics as a new meaningful context for learning the function concept (Stowasser et al., 1994). Computer use has also extended the relevance of "curves" in various directions. For instance, CAD (computer aided design) uses computer based mathematical representations of all kinds of curves and surfaces. In addition, fractal curves have added quite a new visual world to ours.

2.5.2 Functions and the study of motion

The historical emergence of the function concept is intimately related to the study of motion (kinetically and dynamically). Therefore, concepts of calculus and of differential equations were closely related to the new concept of function (Youschkevitch, 1976). These meaningful relations were also in the foreground, when Felix Klein favoured the reform of school mathematics under the banner of functional thinking. The concept of function was seen from the perspective of its meaning in calculus and uses of calculus in the sciences. Interestingly, these relations have been newly evaluated and re-defined in the didactical value systems recently (Kaput, 1993; Kaput, 1994). A newly conceptualised integration of the function concept, the study of motion, and preparatory calculus is being developed under the heading of "the mathematics of change". Time-dependent functions are now considered to be a very important prototype for developing an important element of the meaning of functions and also of the concept of a variable (Freudenthal, 1983; Weigand, 1988). Interestingly, there was a historical controversy about this question whether it makes sense to develop calculus without motion: "in point of intellectual conviction and certainty, the fluxional calculus is decidedly superior [to the French and German versions]; to think of calculus 'without motion' was akin to thinking of 'war without bloodshed, gardening without spades'" (From O. Gregory's 11th edition of C. Hutton's *Course of Mathematics* of 1837; quoted in Howson, 1982, p. 251).

Laws of motion are different from descriptions of motion as time dependent functions. The idea that local causes (forces) "act" at a point to influence the next "step" in a particle's movement is a basic idea underlying differential equations and dynamic systems in general. It is intimately related to the co-variational aspect of functions.

The historical expulsion of "time" from mathematics is challenged by the above suggestions. The current division of labor between disciplines that has brought forth new isles of meaning may not be the relevant separation for mathematics at teacher education and school level. Even if a reunification may be illusionary, teachers should be aware of interfaces, borderlines, and (historical) relationships as a background of their teaching in school.

2.6 Functions as models

If functions are used in a modelling context, all the concepts I have listed under the heading of "applications" and "related concepts in science" become relevant. Science teachers may be better acquainted with these concepts, especially if they have learned scientific research as a process together with some epistemological reflections. For mathematics teachers, however, *proof* as a condition for truth and established knowledge is most important, and the validity of other types of knowledge is difficult for them to judge. I will discuss some aspects of this problem in more detail.

2.6.1 Curve fitting

Curve fitting can be discussed in a purely mathematical context focusing on methods of fitting. Function plotter software has provided new possibilities for doing curve fitting easily. Family of curves described by a parameter set (for instance, family of parabolas) can be used as a repertoire to select from. Geometrical transformations acquire a new relevance in this context, because changing a parameter value into another can be interpreted as geometrical transformation of curves. Systems of equations for unknown parameters are another aspect. In older books for applied mathematics, the distinction whether a function should pass through all points or only "near" the points is basic under the unifying topic of fitting curves to data. Today, courses in mathematics do not necessarily cover these relations and meanings. From the perspective of applied mathematics and the sciences, concepts such as *interpolation* and *extrapolation* and the notion of the *quality of fit* and *range of validity* are important.

Relations to statistical methods (regression, methods of least squares) are also relevant. If a function fits the data well, on what basis can we extrapolate and how far? Teachers should know something about the scientific critique of curve fitting when it is practised without models from which the family of functions can be derived. Nevertheless, such fitted curves can yield excellent predictions (without understanding) in many cases. Even hand-fitted curves may be acceptable for certain purposes, there is no need for complicated fitting methods in every case. They may unjustifiably suggest the application of scientific methods. In sum, many of the above concepts and values do not belong to academic mathematics, but rather to practical mathematics, but they are nevertheless highly relevant for mathematics teachers. What is the domain of validity of extrapolation and interpolation? Do continuous functions describe the "nature" of the relation, or not? Should genuine discrete models be used instead? These are some of the components of the teachers' system of meaning for functions.

Teachers should also know something of the problems of using certain classes of functions for fitting curves: what are the limitations of polynomials? For many, including some software designers, the next "easy" choice beyond linear functions would be quadratic functions. However, polynomials are often not adequate and Splines are preferable. What is the basic idea of Splines? What about Bezier curves that are the underlying curves in many drawing programs?

In the context of curve fitting, geometrical aspects and geometrical classification of functions are acquiring a new meaning. It can be the case that functions having different algebraic representations are nevertheless very near to each other in small intervals and vice versa. Algebraic "near" is different from geometrical "near".

2.6.2 Univalence of functions and modelling

The meaning of univalence as a characterising property of functions is often discussed in relation to distinguishing functions from more general relations. In history, there have been several reasons for giving up the possibility of multivalent symbols such as $\sqrt{\ }$. Also, the curve of a circle is no longer considered as a function because it does not satisfy the "vertical line test". However, curves like the circle can be modelled as functions on a higher level (as mappings from [0, 1] into the plane). Many of the tests with students and teachers Even (1989) refers to are related to the univalence in the above context of meaning. But a further important context is the use of functions in modelling: Here, univalence is an idealisation or model assumption, and there are many cases, where there are several varying values that are associated with one value of the independent variable. Concepts and techniques from statistics are required for modelling in these situations. A deterministic function model (for every x there is exactly one y) has to be epistemologically distinguished from a mixed statistical-functional model where for every x several y are possible and where we can assume a probability distribution for the possible y's. This distribution in general is dependent of the variable x. However, teacher-students have usually not had enough experience in adequate domains of application to appreciate this. The same applies to many didacticians who have done research on the concept of function. In this context, relating functional dependence and correlational dependence adds to the meaning of functions. The *strength* of a relationship is a new perspective in addition to the *form* of a relationship that is expressed by usual functions (Biehler, 1995).

2.7 Various prototypical interpretation

A classification and identification of prototypical ways of interpreting functions (prototypical domains of application) which summarise essential aspects of the meaning (s) of functions would be helpful for meaning development. We can consider Vollrath's (1989) analysis in this perspective. I will add some aspects that are important in the context of modelling and statistics.

Epistemological distinctions should include that functions can be used to express:

- natural laws,
- causal relations,
- constructed relations,
- descriptive relations,
- data reductions.

These distinctions are quite important to avoid misinterpretations. The relation between the quantity and price of a certain article is a constructed relation: it is imposed by fiat (Davis & Hersh, 1980, pp. 70). Using a parabola to describe the path of a cannon ball has the character of a physical (natural) law. Contrary to this use, a parabola used in curve fitting may just provide a data summary of the curvature in a limited interval. Using functions for describing time dependent processes are different from using functions for expressing causal relations: time is not a "cause" for a certain movement. Also, scientists have partly abandoned the concept of causal relation in favour of mere "functional relation" between two quantities (Sierpinska,

1992). This may be due to philosophical reasons but also to simple pragmatical ones: If we have a 1–1 correspondence, we can invert the cause-effect functional relation to infer the "causes" from the effects.

In many statistical applications, functions are used to describe structure in a set of data that cannot be interpreted as a natural law: "Cartesian curve-fitting uses data to determine (comprehend) the structure (curves=laws) governing the universe. Statistical orientation uses curves (regularities) to determine (comprehend) the structure of concrete sets of data—data about phenomena that are important to understand in their own right." (Wainer & Thissen, 1981, p. 195). For instance, the graph in Figure 3 shows the synchronic relation between fuel prices and fuel consumption per inhabitant and per year in various countries of the world.

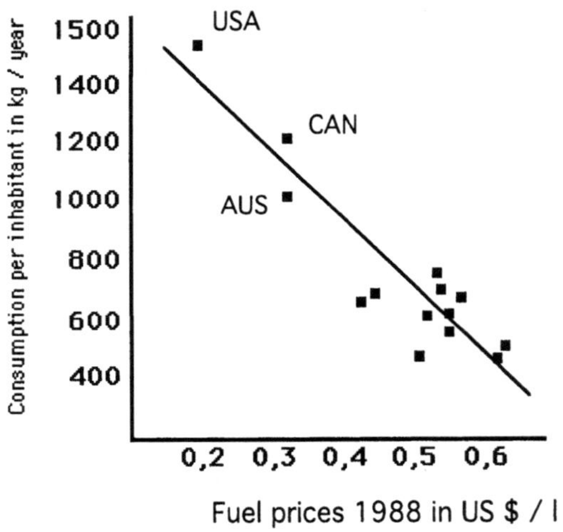

Figure 3: Fuel consumption and fuel prices in various countries
(Data from Weizsäcker, 1992)

If we interpret functions in a causal or natural law sense here, i.e., in the sense of "when we change x, then this results in the following change of y" this will be misleading: we have no direct evidence how the change of fuel price in one country would effect its fuel consumption. We would need diachronic data for that purpose. The above graph can only indicate some evidence. A second remark concerning the above figure: If we exclude North America and Australia from the graph, the rest of the data are only weakly correlated. Statistics requires a very flexible practice of fitting functions to data: excluding points from an analysis or fitting curves only to subsets can be successful tactics. These uses usually are not part of teachers' views of the meaning of functions. Functions are often still taught as if probability and statistics had never been invented.

We will finish our analyses of the meanings of the function concept vis-à-vis teacher education with these remarks. Although a lot has still to be done in doing further research respectively in synthesising research findings under our perspective of meaning reconstruction we hope that we were able to point to important further directions and extensions of current work.

3. SUMMARY AND CONCLUSION

The paper has started with arguing in favour of the thesis that we can re-interpret research and development work in mathematics education as "meaning construction" or "meaning reconstruction". The need is related to the differences between school and academic mathematics, and the situation that school mathematics cannot and should not take over the meaning of concepts in the context of academic mathematics. In the second part, we have looked at the concept of function as an example. A mathematics teacher in-service education course that stimulates enrichment and reorganisation of the meaning teacher-students associate with the notion of function, has provided a concrete context. Relevant but often neglected elements of a meaning landscape of functions have been sketched. The results may help to broaden the background on which we design studies on teachers' knowledge and beliefs about functions.

In addition to this, the paper argued for a systematic approach to the re-construction of the meanings of concepts as an important didactical task. A related research program should aim at knowledge that is less context-bound than knowledge on mathematical meanings that was developed in and for the context of designing concrete curricula and teacher education programs in some concrete reform movement in a very limited period in history.

REFERENCES

Arbeitsgruppe Mathematiklehrerbildung (Ed.) (1981). *Perspektiven für die Ausbildung der Mathematiklehre.* Köln: Aulis.

Balacheff, N. (1993). Artificial intelligence and real teaching. In C. Keitel & K. Ruthven (Eds.), *Learning from computers: Mathematics education and technology* (pp. 131–158). Berlin: Springer.

Biehler, R. (1994). Functional thinking with graphs. Review of "Romberg, T. A., Fennema, E., Carpenter, T. P. (Eds.) (1993), *Integrating research on the graphical representation of functions*, Hillsdale (USA): Lawrence Erlbaum". *Journal of Research in Mathematics Education, 25*(5), 526–533.

Biehler, R. (1995). Funktionales Denken und statistische Zusammenhänge. *Beiträge zum Mathematikunterrich, 1995*, 98–101.

Boero, P. (1992). The crucial role of semantic fields in the development of problem solving skills in the school environment. In J. P. Ponte, J. F. Matos, J. M. Matos & D. Fernandes (Eds.), *Mathematical problem solving and new information technologies: Research in contexts of practice* (pp. 77–91). Berlin: Springer.

Chevallard, Y. (1985). *La transposition didactique: Du savoir savant au savoir enseigneé.* Grenoble: La Penseé Sauvage.

Confrey, J. (1991). The concept of exponential functions. In L. P. Steffe (Ed.), *Epistemological foundations of mathematical experience* (pp. 124–159). New York: Springer.

Confrey, J. & Smith, E. (1994). Exponential functions, rates of change, and the multiplicative unit. *Educational Studies in Mathematics, 26*(2-3), 135–164.

Davis, P. J. & Hersh, R. (1980). *The mathematical experience.* Boston: Birkhäuser.

Dubinsky, E. & Harel, G. (Eds.). (1992). The concept of function: Aspects of pedagogy and epistemology. Washington: Mathematical Association of America, Notes 25.
Ernest, P. (1994). The philosophy of mathematics and the didactics of mathematics. In R. Biehler, R. W. Scholz, R. Sträßer & B. Winkelmann (Eds.), *Didactics of mathematics as a scientific discipline* (pp. 335–349). Dordrecht: Kluwer.
Even, R. (1989). *Prospective secondary mathematics teachers' knowledge and understanding about mathematical functions*. East Lansing, MI (USA), Michigan State University, Ph.D. Dissertation.
Even, R. (1990). Subject matter knowledge for teaching the case of functions. *Educational Studies in Mathematics*, *21*(6), 521–544.
Even, R. (1993). Subject-matter knowledge and pedagogical content knowledge: Prospective secondary teachers and the function concept. *Journal for Research in Mathematics Education*, *24*(2), 94–116.
Freudenthal, H. (1983). *Didactical phenomenology of mathematical structures*. Dordrecht: Reidel.
Harten, G., von, Jahnke, H. N., Mormann, T., Otte, M., Seeger, F., Steinbring, H. & Stellmacher, H. (1986). *Funktionsbegriff und funktionales Denken*. Köln: Aulis.
Höfler, A. (1910). *Didaktik des mathematischen Unterrichts*. Leipzig: Teubner.
Howson, A. G. (1982). *A history of mathematics education in England*. Cambridge: Cambridge University Press.
Jahnke, H. N., Knoche, N. & Otte, M. (Eds.). (1996). *History of mathematics and education: Ideas and experiences*. Göttingen: Vandenhoeck & Ruprecht.
Kaput, J. J. (1993). The urgent need for proleptic research in the represetation of quantitative relationships. In T. A. Romberg, E. Fennema & T. P. Carpenter (Eds.), *Integrating research on the graphical representation of functions* (pp. 279–312). Hillsdale (USA): Lawrence Erlbaum.
Kaput, J. J. (1994). The representational roles of technology in connecting mathematics with authentic experience. In R. Biehler, R. W. Scholz, R. Sträßer & B. Winkelmann (Eds.), *Didactics of mathematics as a scientific discipline* (pp. 379–397). Dordrecht: Kluwer.
Klein, F. (1925a). *Elementarmathematik vom höheren Standpunkte aus II: Geometrie*. Berlin: Julius Springer.
Klein, F. (1925b). *Elementarmathematik vom höheren Standpunkte aus III: Präzisions- und Approximationsmathematik*. Berlin: Julius Springer.
Klein, F. (1933). *Elementarmathematik vom höheren Standpunkte aus I: Arithmetik-Algebra-Analysis*. Berlin: Julius Springer.
Kriesi, C. (1981). *Kognitive Entwicklung durch funktionales Denken*. Bern: Peter Lang.
National Council of Teachers of Mathematics (1989). *Curriculum and evaluation standards for school mathematics*. Reston (USA): National Council of Teachers of Mathematics.
Otte, M. (1984). *Komplementarität*. Bielefeld: IDM Occasional Paper no. 42.
Otte, M. & Steinbring, H. (1977). Probleme der Begriffsentwicklung: Zum Stetigkeitsbegriff. *Didaktik der Mathematik*, *5*(1), 6–25.
Richenhagen, G. (1990). Mathematikgeschichte und Mathematikdidaktik: Überlegungen am Beispiel des Funktionsbegriffs. In H.-G. Steiner (Ed.), *Mathematididaktik Bildungsgeschichte Wissenschafts-geschichte II* (pp. 174–186). Köln: Aulis.
Romberg, T. A., Fennema, E. & Carpenter, T. P. (Eds.). (1993). *Integrating research on the graphical representation of functions*. Hillsdale (USA): Lawrence Erlbaum.
Sfard, A. (1992). Operational origins of mathematical objects and the quandary of reification: The case of function. In G. Harel & E. Dubinsky (Eds.), *The concept of function: Aspects of epistemology and pedagogy* (pp. 59–84). Washington: Mathematical Association of America.
Sierpinska, A. (1992). On understanding the notion of function. In G. Harel & E. Dubinsky (Eds.), *The concept of function: Aspects of epistemology and pedagogy* (pp. 23–58). Washington: Mathematical Association of America.
Steen, L. A. (Ed.) (1990). *On the shoulders of giants*. Washington: National Academy Press.
Steinbring, H. (1994). Reflecting specific characteristics of the object (Gegenstand) of didactical research: Epistemological constraints of mathematical knowledge in social learning settings. Paper presented to the ICME Research Study conference in Washington, May 1994.
Steiner, H.-G. (1969). Aus der Geschichte des Funktionsbegriffs. *Der Mathematikunterricht*, *15*(3), 13–39.
Haapasalo, L.; Stowasser, R.J. (1994). *Computeranimationen. Wiederbelebung der Geometrie.* [Computer animation. Revitalizing geometry] Mathematik Lehren. (Aug 1994) (no.65) pp. 66-69.

Swan, M. (1982). The teaching of functions and graphs. In G. von Barneveld & H. Krabbendam (Eds.), *Proceedings of the workshop on functions* (pp. 151–165). Enschede (The Netherlands): SLO – Foundation for Curriculum Development.

Vergnaud, G. (1990). Epistemology and psychology of mathematics education. In P. Nesher & J. Kilpatrick (Eds.), *Mathematics and cognition: A research synthesis by the International Group for the Psychology of Mathematics Education* (pp. 14–30). Cambridge: Cambridge University Press.

Vollrath, H.-J. (1989). Funktionales Denken. *Journal für Mathematik-Didaktik, 10*(1), 3–37.

Wainer, H. & Thissen, D. (1981). Graphical data analysis. In M. R. Rosenzweig & L. W. Porter (Eds.), *Annual Review of Psychology Vol. 32* (pp. 191-241). Palo Alto, CA: Annual Reviews Inc.

Weigand, H.-G. (1988). Zur Bedeutung von Zeitfunktionen für den Mathematikunterricht. *Journal für Mathematikdidaktik, 9*(1), 55–86.

Weizsäcker, E. U. (1992). Energie und Umwelt. *WIFO Journal, 92*(9), 392–396.

Weth, T. (1993). *Zum Verständnis des Kurvenbegriffs im Mathematikunterricht.* Hildesheim: Franzbecker.

Weyl, H. (1918). *Das Kontinuum; kritische Untersuchungen über die Grundlagen der Analysis.* Leipzig: Veit. [Reprint: Das Kontinuum und andere Monographien. New York: Chelsea 1960]. English translation: Weyl, H. (1994). *The continuum: a critical examination of the foundation of analysis.* New York: Dover Publications.

Williams, S. R. (1993). Some common themes and uncommon directions. In T. A. Romberg, E. Fennema & T. P. Carpenter (Eds.), *Integrating research on the graphical representation of functions* (pp. 313–337). Hillsdale (USA): Lawrence Erlbaum.

Winkelmann, B. (1988). Funktionskonzepte in der Interaktion zwischen Benutzer/Lernenden und Rechner in mathematischer Unterrichts- und Anwendersoftware. *Zentralblatt für Didaktik der Mathematik, 20*(5), 222–228.

Youschkevitch, A. P. (1976). The concept of function up to the middle of the 19th century. *Archive for History of Exact Sciences, 16(1)*, 39–85.

Zbiek, R. M. (1992). *Understanding of function, proof and mathematical modelling in the presence of mathematical computing tools: Prospective secondary school mathematics teachers and their strategies and connections.* University Park, PA (USA), Pennsylvania State University, Ph D Dissertation.

OLE SKOVSMOSE

MEANING IN MATHEMATICS EDUCATION

Educators would certainly agree that mathematics education should be meaningful to the student—and I am no exception. But what could the meaning of *meaningful* be? Here one finds a variety of opinions.

One attempt to insert meaning into mathematics education took place as part of the curriculum reform movement in the 1960s. The drill and practice of traditional mathematics teaching was to be replaced by real understanding, which was interpreted as understanding logical relationships between mathematical terms. Meaning was to be established in terms of "logical honesty". If students were enabled to see structural connections behind algorithms, then more meaning would be brought into the classroom.

Meaning can also be described in relation to social structures, which requires that the whole educational process be taken into consideration. To introduce students to the culture of mathematics means an enculturation. And in order to ensure successful enculturation, the teacher must know about the culture to which the students are introduced as well as that from which they come. It is essential, therefore, to relate the content of the educational process to the students' background. I see the definition of *meaning* in relation to the cultural background of the children as a huge improvement on the definition of meaning as related to logical structures. Nevertheless, I shall also emphasise some shortcomings of this way of pointing out meaning in education.

My book *Towards a Philosophy of Critical Mathematics Education* (Skovsmose, 1994) contains descriptions of a few examples of project work in mathematics education carried out in secondary schools (for other examples, see Nielsen, Patronis & Skovsmose, 1999; Alrø & Skovsmose, 2002). These descriptions served to clarify some of those concepts by whose means I sought the goal set out by the title of the book. In that book I only touched briefly on the notion of meaning, although the whole study can be interpreted as a concern for meaning in mathematics education. In this chapter, I reinterpret some part of my work, and with reference to two of my examples, I try to make more explicit the notion of meaning.

1. DIFFERENT INTERPRETATIONS OF MEANING

Philosophers have given a variety of interpretations of the concept of meaning. In the first paragraph of *Philosophical Investigations*, Ludwig Wittgenstein (1953) quotes Augustine, who interpreted meaning in terms of reference: A word has a certain meaning because it refers to a certain entity. A related theory is expressed by Wittgenstein (1961) in the *Tractatus Logico-Philosophicus,* in which language is clarified as a "picture" of reality. This concern about reference is also part of the work of Gottlob Frege (1969), who refined his philosophy of meaning into a theory of "sense" and "reference". Along with the interpretation of meaning in terms of reference is the assumption that the meaning of a composite expression is determined by the meanings of its linguistic elements. In other words, the meaning of a molecular expression is a function of the meanings of its atomic constituents.

In *Philosophical Investigations,* however, Wittgenstein (1953) criticises the referential theory of meaning and suggests that it should be replaced by a different understanding of meaning. He finds that the meaning of a sentence must be related to the use it is possible to make of the sentence. Maybe one can even identify the "use of the sentence" with the "meaning of the sentence". This identification opens a new horizon in the philosophy of meaning. Now the meaning of the sentence can be related to the complexity of the whole situation in which the sentence is used. The meaning of a sentence has to be understood in "the stream of life".

Instead of looking for the meaning of a word, one looks for the meaning of a linguistic act. This interpretation, however, introduces another alteration. If the meaning of a sentence can be interpreted as the use of the sentence, then "use" and the "context of use" become semantic concepts. Why pay special attention to the notion of a sentence? To look for the meaning of (the use of) a sentence is no longer the pre-eminent choice. One might as well look for the meaning of (the use of) a formulation, a gesture, a text, an attitude, or any other action.

This broader interpretation is further developed by Jürgen Habermas (1984) in *The Theory of Communicative Action* (especially in the chapter "Intermediate Reflections: Social Action, Purpose Activity, and Communication"). In order to define *illocutionary meaning,* Habermas draws upon the speech act philosophy of language initiated by Wittgenstein (1953) and by John L. Austin (1946, 1962) and further developed by John R. Searle (1969, 1971). What is done when a speech act is performed? To answer that question, the *locutionary content, illocutionary force,* and *perlocutionary effect* of the utterance are investigated. Meaning comes to refer to the practice, the context, and the commitment of the persons who take part in the communicative action. To understand meaning, therefore, presupposes a view not only of the person who expresses a statement but also of the whole situation in which the communicative action takes place.

This approach to the discussion of meaning can also be illustrated by a reference to the work of Anthony Giddens. In *The Constitutions of Society*, Giddens (1984) describes a theory of structuration. He emphasises that "human action" plays a basic role in sociology, and he rejects empiricism and structuralism, which have assumed that "sociological facts" constitute the ultimate object of sociological studies. When, instead, "human action" becomes the focus of sociology, the notion of "fact" becomes blurred. "Human action" is not an adding up of individual "acts": "'Actions' is not a combination of 'acts'" (p. 3). Action takes place in the *durée* of lived experience, and this *durée* is significant for grasping the meaning of an action (p. 3).

"Practice" is a general and muddy notion; but in an attempt to specify, one can introduce the notion of *sphere of practice* used by Pierre Bourdieu.[1] A sphere of practice makes it possible to interpret different acts (which an "outsider" might have identified analytically) as being part of a lived reality that provides meaning to the different acts. Analytically isolated, or observed as unique and significant, a specific act might appear to be without meaning, but located in a sphere of practice it makes sense. The sphere of practice thus becomes an important unit for understanding the activity in question. Instead of "sphere of practice", one can also talk about a "network of tasks" or a *durée* of lived experience.

2. MEANING AS AN EDUCATIONAL CONCEPT

It is possible to ask about the meaning of a mathematical concept. It is also possible to ask about the meaning of a (mathematical) task as part of an educational practice. Both questions express a concern for bringing meaning into the mathematical classroom. But the two questions are different. My fundamental assumption is that for students to ascribe meaning to concepts that have to be learned, it is essential to provide meaning to the educational situation in which the students are involved. The meaning of concepts does not provide an adequate foundation for the meaning of tasks. What might a concentration on "meaning of task" mean to mathematics education?

A preoccupation with references has played an essential role in the conception of meaning in the set-theoretical approach to school mathematics. To understand the meaning of *function,* students must understand the meaning of *Cartesian product, subset,* and so forth. According to this approach, learning progress can be discussed in terms of the semantic landscape through which the students are travelling. It is usually recommended that the students start their journey in an empirical landscape. Functions, as well as all other mathematical concepts, get a physical interpretation; a function becomes a certain kind of machinery, and so on. Then the travel can continue into more abstract "semantic fields", and mathematical notions come to refer to abstract entities. This educational approach has become further developed into a set of priorities for research in mathematics education.

Inspired by constructivism, the main question has become "What does this concept allow the person to do?" instead of "What is the reference of this concept?" Concepts are not delivered, they are constructed. The meaning of a concept can, therefore, be associated with what the person can do by means of the concept.

1 My use of *sphere of practice* is inspired by the discussion taking place in the BACOMET group. I do not try to develop the notion with reference to the work of Bourdieu.

"Meaning" and "viability" become connected concepts. This line of thought, however, still concentrates the discussion of meaning in mathematics education on the meaning of (mathematical) concepts.[2]

Important questions become the following: What sort of meaning can be associated with certain mathematical concepts? What is the meaning of a particular concept to the students? What sort of meaning can be associated with this concept from a mathematical point of view? What is the meaning of this concept from the perspective of the teacher? What is the shared meaning of this concept (an important question when social constructivism is considered)? The priorities incorporated in these questions make up a paradigm in the sense that they establish preferences about the object of research in mathematics education. This paradigm I label *conceptism* because the meaning of concepts gets first priority when meaning in mathematics education is discussed. (Conceptism can also be seen as a holdover from the assumption that meaning of a composite expression is a function of the meanings of its atomic constituents.) If a conceptual framework, as elaborated by Willibald Dörfler (1991), is used for a discussion of meaning, one witnesses conceptism. Many studies of mathematics learning and of classroom communication, not least those inspired by radical constructivism, concentrate on the meaning of concepts.[3]

It is not a straightforward empirical question whether the meaning of concepts takes priority over the meaning of tasks. The prioritisation of the different notions of meaning is therefore "paradigmatic". As I see it, conceptism establishes a research programme that tends to eliminate essential educational issues. It seems to provide a particular discourse that, in my opinion, contains "blind spots". In particular, the socio-political and cultural context of the students will easily appear insignificant for the study of meaning in mathematics education. But if the *durée* of lived experience is significant for meaning construction, then the meaning of concepts might provide a poor start for a discussing meaning.[4]

The key idea of this chapter is that the discussion of meaning in mathematics education cannot be structured by the priorities of conceptism. Instead, the basic discussion of meaning has to do with the meaning of the activities in which students are involved as part of an educational process. Only then, as a specification of this

[2] A more advanced theory of meaning, still building upon the referential paradigm, is expressed by the triangle (referent, reference, symbol) developed by C. K. Ogden and I. A. Richards and presented in 1923 in *The Meaning of Meaning*. The work of Ogden and Richards corresponds nicely with C. S. Peirce's triangle (sign, object, interpretant). (For an introduction and full references, see Fiske, 1988.) This model has been further elaborated to form a basis for discussing meaning in mathematics education.

[3] See, for instance, Cobb & Bauersfeld (1995); Noss & Hoyles (1996); Seeger, Voigt & Waschescio (1998). Another illustration of what conceptism might or might not mean can be found in the index of the *Handbook of International Research in Mathematics Education*, edited by Lyn D. English (2002), under the following entries: "meaning", "meaning in learning", and "meaning of mathematical propositions".

[4] For example, much research in using the computers in the classroom suffers from the limited perspective of meaning in education supported by conceptism. This phenomenon was emphasised by Keitel, Kotzmann & Skovsmose (1993). See also Skovsmose & Valero (2002a).

discussion, can the meaning of mathematical concepts be investigated. I do not, of course, claim that investigations of meaning of concepts are irrelevant. My critique is directed towards the claim that investigations of the meaning of concepts deserve first priority in an educational investigation of meaning. Meaning has to be discussed in terms of spheres of practice in which students are involved, which involves a paradigmatic shift away from conceptism.[5] One way of dealing with this shift is to share the concern of critical mathematics education (Skovsmose & Nielsen, 1996; see also Skovsmose, 1998a, 1998b, 2000, 2002a; and Skovsmose & Valero, 2001, 2002a). This possibility, however, need not be the only one, and in what follows I restrict myself to a more general indication of an alternative to conceptism.

3. A NOTE ABOUT AN EMPIRICAL STUDY

One of the projects described in *Towards a Philosophy of Critical Education* (Skovsmose, 1994), "Family Support in a Micro-Society", concerns child benefits.[6] First, the students (14–15 years old) were asked to invent a micro-society. This task involved the description of 24 families, their income, the number of children in the family, and whatever additional information was considered interesting. The descriptions were put together in a small magazine referred to as the "Family Circle", which then constituted the micro-society. Then the students were split into five groups, and each group was to conceive of themselves as being the political representatives of a district. Their first political task was to set up guidelines for distributing child benefits in their district. Next, a certain amount of money was given to the districts, and the students had to establish a mathematical algorithm for the assignment of child benefits based on their general "political" guidelines for such a distribution. Having finished this administrative task, they compared the results of the different districts, and the acceptability and political fairness of the different distributions were discussed.

In the project "Energy", students discussed the input-output figures for "use of energy".[7] The first part related to the students' own breakfasts. What energy supply does a normal breakfast contain? This energy supply was calculated by using

5 My concern is to establish a mathematics education that views the students' experience as meaningful. Therefore, I refer to the students' meaning. The analysis, however, could naturally have been developed in terms of meaning seen from the perspective of the teacher, the perspective of society (maybe interpreting schooling as a preparation for participating in democratic life), or the perspective of a future workplace.

6 The project took place at Klarup Skole, and the teacher was Henning Bødtkjer. For a description in greater detail, see Skovsmose (1994, ch. 9). I do not intend the example to illustrate what "ought" to be done in mathematics education. The example serves not as a prescription but as a reference point for what I want to say about "meaning". I do not hesitate to say, however, that Bødtkjer developed most valuable and interesting examples.

7 This project was also carried out at Klarup Skole with Henning Bødtkjer as the teacher (see Skovsmose, 1994, ch. 7). I use "use of energy" as in everyday language. Physics states that energy does not disappear but changes from one form to another. It is this phenomenon of changing, of course, that is referred to in the expression "use of energy".

statistics about the "energy content" of bread, butter, cheese, and so forth. "Use of energy" consisted of a trip on a bicycle. By means of certain formulas involving the parameters of velocity, time and "front area of the cyclist", the use of energy during the trip was calculated. On this basis, an input-output account could be set up with reference to the breakfast and the cycling. Then the project turned to input-output figures for farming. First, how great an input of energy is needed to grow barley in a certain field? This input includes, for instance, the use of petrol for ploughing. Next, the students calculated the energy supply contained in the harvested barley. The result of these calculations showed that the energy output of the barley was six times the energy input: a promising input-output account! The farmer, however, used the harvested barley as pig food, and the input-output figures for pig breeding were then calculated. The result was that the energy output was one fifth the energy input. According to the students' calculations, therefore, pork production is very expensive seen from the perspective of energy supply. These results were then discussed in a global perspective.

As part of the project "Energy", it made sense for the students to ask: "What does "function" mean? What does it mean to solve an equation?" The point, however, is that it also makes sense for the students to ask questions like: "What do input-output figures in farming refer to? Why should we carry out these calculations? Will we get some reliable results? What, in fact, do we learn from all this?" Mathematics education is often framed by the discourse of exercises, and in this case, it makes sense to ask the first type of question. However, it hardly makes sense for the students to consider questions dealing with the purpose of their activities. (The "logic" of the discourse in the exercise signals that the purpose of doing exercises is simply to get on to the next exercise.)

To make mathematics education meaningful, I find it essential to establish an organisation that invites students to discuss the meaning of their different tasks. The purpose of doing something should be made available to the students in a language that is also accessible to them. This discussion is important to me as an educator. As a researcher it is important for me to suggest a framework for research that does not obstruct the possibilities for investigating meaning from this broader point of view. The investigations of the meaning of concepts are not sufficient for an investigation of meanings that students might assign to their tasks.[8]

The two examples, "Family Support in a Micro-Society" and "Energy", can be seen as attempts to contextualise mathematical activities in such a way that these activities can be grasped as meaningful by the students. Naturally, the contextualisation need not be accepted by the students. They need not be motivated. Still, the contextualisation is an attempt to make it possible for students to negotiate the meaning of the tasks in which they are involved.

[8] When one discusses meaning in educational practice, one always has to face a basic contradiction that differentiates the school situation from many other situations. Research by Christiansen (1994) points directly to the phenomenon that the "school setting" can corrupt a well-intentioned contextualisation. This observation naturally applies to "Energy" and "Family Support in a Micro-Society" as well.

4. DISPOSITION, INTENTION AND ACTION

If meaning in mathematics education is to be related to the meaning of tasks and not primarily to the meaning of concepts, a specification of the notion of learning is needed.[9] First, I take a closer look at the notion of "action". I try to describe "meaning in education" in terms of "meaning of action" by interpreting (some) learning as action. I do not suggest that all forms of learning can be seen as action. It is possible to be forced to learn something, and some learning can also take place beneath one's level of awareness—for instance, when one assimilates certain habits. For me, however, learning as action represents an important form of learning, as it can turn into critical learning (for a discussion of the relationship between dialogic learning and critical learning, see Alrø & Skovsmose, 2002).

In my terminology, we cannot say that a person is acting and, at the same time, that he or she is being forced to do what is actually done. The acting person must be in a situation where choice is possible. The person acting must also have some idea about goals and reasons for obtaining them.[10] This property differentiates action from "blind activity", like automatically biting a pencil when the next sentence becomes too difficult to formulate. *Action,* as I use the term, presupposes a degree of indeterminism and a degree of awareness.

This usage connects action with *intention*. Actions cannot be described in purely mechanical terms. Such a reduction was once suggested, but I see action as categorically different from mechanical behaviour. This difference is expressed by the fact that intentions are important in the identification of an action. We cannot describe an action without describing the orientation of an individual. To ask whether a person's intentions are fulfilled corresponds to asking if he or she has performed certain actions. The intentions represent, so to speak, the (personal) meaning of the action. An action does not, however, consist merely of an initial intention expressed before the actual activity takes place, and then the activity itself. One must also talk about intentions within the action.

Intentions do not spring to life from nothing. They are grounded in a landscape of pre-intentions or *dispositions*, and I divide these into "background" and "foreground". The background of a person can be interpreted as the socially constructed network of relationships belonging to the history of the social group to which the person belongs.[11] When one tries to understand an individual's intentions, one often refers to her or his background. But equally important is the person's foreground. By this, I refer to those opportunities that the social situation makes available to the social group to which the person belongs. Opportunities are not to be understood as sociological facts but as collectively or individually interpreted opportunities. Dispositions are objectively rooted, but they are not factual.

9 If the perspective had been that of the teachers, then the focus should have been "teaching" and not "learning". What follows summarises the discussion in Skovsmose (1994, ch. 10).

10 This statement has to be interpreted in a broad sense because a person may well be acting even if the goal is hazy and if the reason for obtaining that goal is obscure or merely implicit in a situation.

11 For simplicity, I ignore the fact that a person might, depending on circumstances, belong to different social groups.

Dispositions are mediated by the individual, and they express both a subjectivity and an objectivity.

Dispositions are just "dispositions", which means that they cannot be observed directly. They signify propensities. Dispositions can, however, be revealed when a person acts. It does not make sense to talk about dispositions as the cause of intentions. Intentions emanate from the individual's background as well as foreground. The individual produces (or raises, or creates, or makes decisions about) his or her intentions and in so doing, reveals the dispositions in his or her actions.

Actions have effects, and it makes sense to try to interpret the concept of action also in terms of the person's reaction to these effects. This interpretation opens up a cyclical process: Dispositions become changed because of reflections on intentions and actions, and on consequences of these actions. Dispositions are grounded in the individual's social objectivity and are simultaneously produced by the individual, partly as a consequence of the actions he or she performs. The success or failure of actions gives rise to modified dispositions. Through actions, dispositions become moulded and thereby become the source of new intentions.[12]

I see "meaning of action" (or "meaning of task") as referring to a pattern including: (a) the dispositions of the person, including both foreground and background; (b) the person's intentions; (c) the person's actions, including the intended as well as the unintended effects; (d) the person's reflection on these effects; and (e) the feedback of these observations on the person's dispositions. Therefore, to discuss the meaning of an action, one has to consider the entire situation in which the action takes place. In particular, a sphere of practice and a *durée* of lived experience can be discussed in terms of disposition, background, foreground, intention, and reflection. Naturally, I do not claim that these concepts constitute a sufficient conceptual framework for specifying a sphere of practice. I claim only that these concepts are relevant when the notion of meaning is discussed with reference to a practice.

5. LEARNING AS ACTION

Some processes of *learning* can be interpreted as similar to processes of action (Skovsmose, 1994; see also Alrø & Skovsmose, 2002), which brings the concepts of disposition, background, foreground, intention, and reflection into use for defining a concept of learning.

Dispositions constitute a totality from which intentions of actions emerge, and therefore when one discusses learning, one is concerned with "background" as well as "foreground". The situation that can raise intentions of learning does not automatically belong to the student's background, in terms of her or his situation and social heritage. It also depends upon the student's opportunities in future life as the

[12] I have outlined this circle, however, in a way that needs strong modification. I have talked about actions being undertaken by an individual, but it might be better to see actions of a group as the principal conceptual unit, which would mean that co-action or co-work would be the primary unit instead of action or work.

student perceives them. Therefore, as mentioned previously, I am not so happy with those approaches in mathematics education, like certain examples of ethnomathematical studies, that seem to refer the discussion of meaning to the students' background (see, e.g., Skovsmose, 2002b).

Intentions of learning may emerge out of dispositions. The decisions of the learner, therefore, play a role when conditions for learning are produced. I do not see intentions of learning as being different from other sorts of intentions except that they can be fulfilled by learning activities.[13] Besides talking of intentions of learning as initiating a learning activity, one can talk about intentions in learning. The student has to be *involved* in the learning if the learning activity is to become learning as action. Intentions *of* and *in* learning must exist. A learner can learn many things by command, but if learning means not just receiving information but also includes reflection and a critical awareness, the learning has to be *performed* by the learner.

Students reflect on what happens in school. These reflections need not, however, be formulated and expressed in terms that are in accord with school practice as intended by the teacher. Students have experiences in school related to the content, but certainly also related to the teacher, their classmates, and so on. This whole network of experience is reflected on by students. No conclusions need to be formulated, but still the reflections can change their disposition. Students can alter their plans for what they want to be—they can see their future in a different light—and that can become condensed into dispositions. Therefore the students' intentions also express conclusions resulting from their reflections.

Formulations such as "learning must be performed by the learner" do not mean "voluntarism" in the sense that every possibility is open to the person if only he or she decides to act. An action is something that has to be performed, and that performance presupposes intentions that can be raised from the person's dispositions. But the person is not in control of her or his dispositions. They are socially structured and, at the same time, interpreted.

We cannot, as teachers or curriculum designers, implant goals in a student, nor can we implant good reasons. Goals have to be identified and accepted by the learner. Reasons have to be accepted; if not, they can never become the person's reasons. But they must not only be accepted, because the intentional orientation must be performed by the persons themselves. A condition for a productive teaching-learning process is that a situation is established in which students are given opportunities to investigate reasons and goals for suggested teaching-learning processes, and by so doing, to accentuate their own intentions and to incorporate some of them as part of their learning processes (see Vithal, 2003).

When learning is interpreted as action, the notion of *frame* becomes important. "Frame of learning" also means "frame of action". By a frame I understand an

[13] It might be useful to distinguish between a learning activity and the accomplished learning seen as the result of the activity. One might consider whether it is the learning activity or the accomplished learning that should be seen as the condition for fulfilling the intentions of the learning. In my interpretation, it is the learning activity, although serious objections could be raised about that interpretation.

educational structuration that highlights a part of the students' reality as a source for a teaching-learning process and as a resource for establishing reasons for learning. "Frame" refers to all such different ways of establishing a "learning milieu" that can help make the students' tasks meaningful. I see a thematic approach (as exemplified by "Energy" and "Family Support in a Micro-Society") and project work as attempts to establish frames in education.[14] A concern for framing learning activities indicates a perception of students as acting in their learning. Without a frame, the educational process in mathematics is pushed forward by the inertia of the (nearly) infinite sequence of exercises, in the sense that the rationale for solving one exercise is simply to get on to the next one. The teacher and the textbook take control of the process, and the students cannot be acting persons.

In "Energy" and "Family Support in a Micro-Society", the framing made it possible for the students to see a suggested purpose of what they were doing, and they were enabled to answer a question like: "Why are you doing these calculations?" To frame an educational process means to consider the students dispositions as being raw material for reasons of learning. They can make decisions on what to do; they can decide to do some calculations and omit others. Framing in education is an attempt to impart meaning to the educational process. This attempt, naturally, does not guarantee successful education, but that is not my point. My point is simply that a successful framing invites students to *act* as learners—and naturally they can decide not to undertake the project. My claim is not that framing provides educational success but simply that one should see framing as recognition of students as acting persons.

Inspired by conceptism, bringing meaning into education has come to mean introducing the different mathematical concepts with a heavy reference to daily life situations. For instance, *function* might be illustrated with all sorts of linkages between two parameters. Naturally, I do not claim that such specific examples are misleading. What I want to emphasise is that even if all mathematical concepts are introduced during the course with a garnish of practical examples, the whole process of education may still appear as highly abstract as long as the purpose of the process is not rooted in the horizons of the students. The essential thing is to bring meaning to the educational actions of the students—and that meaning is not the sum of the meanings of different concepts.

6. (DIFFERENT) MEANINGS IN LEARNING

Like the meaning of action, meaning in learning comes to refer to a relationship between the dispositions of the learner, the intentions of the learner, the intended and unintended effects of learning activities, and the learner's reflections on these

[14] The work of Freire (1972, 1974) illustrates framing, as situations from daily life are identified as starting points for an educational programme. For a discussion of project work in mathematics at university level, see Vithal, Christiansen & Skovsmose (1995). See also Skovsmose (1994) for a discussion of "scene setting" and Skovsmose (2001) for a discussion of "landscape of investigation".

effects. That means the issues to be considered, when meaning in mathematics education is discussed, become very complex.

It does not make sense to talk about intentions of students as something pre-existing even though the students' dispositions are resources for intentions. Intentions of learning can unfold in many ways: they can be advanced, refined, restructured, discharged, or scrapped. And therefore the meaning, as interpreted by the students, can vary enormously even though the specific task seems to be the same.

The emergence of intentions for learning often takes place in a situation overburdened by demands. A mathematical textbook is usually a carefully elaborated sequence of commands, reflecting the directives put forward in the curriculum and verbalised by the teacher. The structure of schooling makes a forceful structuring of the pre-intentional dispositions. In a normal classroom situation, intentions of learning are rarely seen emerging as part of a negotiation in which the teacher expresses possibilities and the students express themselves in order to grasp the situation better. Nevertheless, the activity of changing and adjusting intentions is a very common activity in school. The demands of the situation make it necessary for the students to restructure their intentions, but that often happens in a haphazard way. The adjustment of intentions does not take place as a shared enterprise but as individual undertakings. A multiplicity of different intentions, not necessarily having much to do with learning, may be set up. The meaning that the individual student could add to the activity is, so to speak, out of control. Ascription of meaning is individualised.

The result is not that no learning takes place but that learning may congeal as a *forced activity*. The demands of the situation influence the intentions that the students may add to the individual learning activities. The intentions in action become strategic. Because students interpret the school situation in an ongoing way, the demands of the situation become part of the students' dispositions and, therefore, part of the situation that students consider when they determine how to act. Students will enter school with ideas, hopes, and expectations. But the demands of the situation in school too often result in *broken* or *ignored intentions*. When students' intentions are ignored, it seems impossible for students to perform actions that can fulfil negotiated intentions. That can happen when students are not given any possibility to express goals and reasons, or it can happen when they are not offered any reasons for what is going on.

The development of *underground intentions* is a common phenomenon in classrooms. These intentions are not shared in the "classroom public". When the intentions of the students are ignored, the consequence is not that the students are emptied of all sorts of intentions and prepared to accept the delineated aims and reasons. Instead, an abundance of underground intentions will emerge, and the student's curriculum will easily rise to confront the official curriculum.[15] Students

[15] Lindenskov (1993) discusses the notion of the student's curriculum. Alrø & Skovsmose (2002, ch. 5) illustrate how underground intentions can unfold in the classroom situation and turn into a flat refusal to learn.

cannot act as part of a common learning process; they become stripped of power, but they create a new space for possible actions. Each student can invent a great variety of underground intentions that give meaning to life in school when the curriculum itself has become stripped of meaning.

This phenomenon indicates that some sort of framing will always take place. Whatever happens, the students will interpret the situation. Students may be prevented from being involved in the educational process as genuine participants, but they can develop underground intentions. They can act in the classroom with reference to a frame that does not belong to the school public. Framing, as in "Energy", refers to a deliberate attempt to bring part of the students' horizon into the classroom to serve as a reference for learning activities. But students can always do their private framing.

In fact, it makes sense to talk about multiple framing.[16] The school structure provides a frame. For instance, students cannot ignore that marks have to be given and that tests will follow. The students observe such events. The teacher might suggest a theme, say, "Energy". But several conflicts are possible. A student may want to ask a question that seems highly relevant with regard to the project. But the question can also be interpreted in the context of the class community, and maybe the student does not want to reveal to his or her classmates or to the teacher that he or she has to ask this question. The project still takes place in a situation where marks have to be given. Whatever happens, the students will interpret the situation. In this sense, framing will always take place.

Broken intentions are not the same as *modified intentions* or *integrated intentions* or *shared intentions*. The teacher has ideas and plans, the students likewise, and no parallelism can be presupposed. The sharing of intentions presupposes a complicated procedure involving the imagining of different goals and reasons. The relevance of dialogue and negotiation has precisely to do with this modification. By means of the activity of sharing intentions, the students may come to act as a group and add a new force to the dynamics of learning. An educational task is to introduce a frame that can support reasons for learning, with due recognition of the fact that students may act in terms of quite different intentions.

As intention connects with meaning, the variety of intentions that might occur in an educational situation gives rise to a variety of meanings that the students might ascribe to their activities. Thus students might calculate the value of a function (giving meaning to a mathematical notion); they might apply the function to determine the energy supply of a certain meal (giving meaning to an application of mathematics); they might find the result "suspicious" (giving meaning to a reflection on an application of mathematics); they might try to calculate the value of the function because other students are doing the same (implying that the meaning of their actions can only be discussed in terms of instrumentalism); or they might simply find doing calculations interesting (importing an intrinsic meaning to

[16] Christiansen (1994) discusses the influence of the school setting on students' interpretations of their educational tasks. This study is essential for the further investigation of "interferences" between the different framings.

mathematics). A student might engage in the calculation because he or she wants to cooperate with a certain classmate, and so on. The meaning of a learning activity can only be understood in a space of multiply-oriented intentions. Research dealing with meaning in mathematics education must try to grasp this complexity.[17]

This conclusion can be stated in a different way. When one claims that a certain mathematics learning environment gives mathematical concepts a deeper meaning (whatever the interpretation of *deeper* might be), that claim becomes dubious when it is not discussed within a framework that recognises the space of multiply-oriented intentions in which learning takes place.[18]

7. FURTHER COMMENTS ON THE PROJECTS

The formulas used in the project "Energy" for breakfast-cycling input-output had been obtained from sports medicine sources; they were simplifications of the formulas actually used when calculating air resistance for a moving bicycle. The students knew the source of the formulas, which provided the formulas with an authority not normally found with the artificial formulas most often used in mathematics teaching. The students' attitudes towards these "reliable" formulas were quite different from their attitudes towards invented formulas. The formulas for air resistance were discussed and their plausibility evaluated, even though the students knew them to be grounded in research.

The critique of mathematical formulas does not simply spring from an elaborated and well-grounded interpretation of the meaning of concepts. That critique may stem from a stratum of meaning, which has to do with the meaning of the activity itself. The concern for the reliability of the formulas that expressed air resistance would not easily have been established if everything had been developed in accordance with the exercise paradigms common in mathematics education. The "Energy" project could easily be transformed into a sequence of exercises. The same formulas could have occurred, including the same values of the parameters. One could indeed imagine that, in the reformulation of "Energy" into a normal sequence of lessons, the exemplification of the different formulas was decorated with a superabundance of examples. Every mathematical term could have been embodied in the daily experiences of the students. Nevertheless, the educational situation would be quite different. The framing actually used brought the students into the process and, in particular, brought the importance of critique into the classroom. And a source of critique is to be found in the broader meaning of the activity.

Different types of terminology were used during the project. Mathematical expressions were developed by means of which the calculations were conducted. A

17 Attempting to grasp such complexity can naturally become an overwhelming task. My point, however, is that the task should not be simplified because of a paradigmatic restriction. The research of meaning into mathematics education should acknowledge all aspects of meaning.

18 A further elaboration of the notion is found in Alrø & Skovsmose (2002), where the notions of "dialogue", "intention", "reflection" and "critique" are brought together in a theory of learning mathematics that resonates with critical mathematics education.

second type of terminology was developed when the discussion of these formulas took place: What are the general ideas behind the formulas? This talk about formulas became less abstract than might have been expected and, in fact, quite understandable to students unable to do the calculations on their own. The reason seems to be that they had gained a personal feeling for what the formulas might mean. That may seem surprising because investigating formulas in general terms is normally thought of as an abstract task. But the students were able to interpret the general ideas behind the formulas. The ability to carry out calculations seems not to be a necessary condition for understanding the "jobs of formulas" when students have some informal experience of what the formulas express. The cardinal point here is that the general explanation given by the teacher was made in terms understandable to (perhaps) all the students. The input-output terminology had become quite commonplace. Even though several students were unable to perform the calculations on their own, they were still able to understand what the calculations were about—and eager to know about the final results. The framing provided meaning to the different levels of terminology.

During the educational process some "vantage points" were established from which it was possible for the students to "see" what had been achieved and what was going to be done during the next phase of the project. The direction of the educational process and the purpose of the different tasks did not have to be expressed in mathematical or technological terms. When a vantage point is established, it is not an educational catastrophe if students do not understand the details of the calculations that follow. They still have a meta-conception of what they are doing. To calculate the use of energy in barley farming, the relevant information was the width of the machines used and the number of times the field had to be gone over, and *not* a specific perception of the size of a one-hectare field, which, in fact, had been measured out when the students and the teacher arrived at the farm. From a mathematical point of view, the actual measuring of one hectare was quite unnecessary. But the measuring of the field was essential for providing a point of reference for later discussion. During the project, several references were made to this piece of land, and it was possible to provide summaries characterised by a well-established concreteness: that the calculated input-output figures show that the proportion one to six means that the farmer can harvest barley containing about six times the energy he has to use when harvesting, sowing, ploughing, and whatever else has to be carried out in the area measured. The terminology used for explaining the general ideas of the calculations is provided with specific meaning: The semantics for the discussion of the educational process were enriched. Measuring the field established a meaningful vantage point.

One of the main purposes of framing is to produce vantage points. Vantage points become hills in the semantic landscape of the educational process. A vantage point is established within the students' horizon, and from this point it is possible to "survey" the different tasks. The vantage point helps provide students with a meta-language about the different tasks in the educational process. Vantage points become semantic conditions for communication between students and teacher, not only about the strictly educational content but also about the perspectives of this content as well as about what is done and what has to be done in the classroom. Vantage points help to establish a condition for negotiating the "meaning of tasks".

8. CONCLUDING CONSIDERATIONS

Dispositions, intentions, and reflections are essential in order to understand "action" and, therefore, "learning as action". I have described dispositions as constituted by both the background and the foreground of the students. The notion of foreground brings us more directly to the political dimension of "meaning production" (Lins, 2001).

The foreground is the set of opportunities that the student's interpretation of his or her socially determined opportunities reveals as "real" opportunities. In this way, a foreground is a subjectively mediated, socially determined fact. The foreground of the person acting is an important source for understanding an action. Similarly, the foreground of a learning person is an important parameter in understanding the learning process and the meaning that the students might ascribe to that process.

To concentrate solely on the background as a source for understanding "meaning in mathematics education" is problematic. In a simplified version of ethnomathematics, it has been suggested that to make mathematics education meaningful, the content of the education should reflect the students' background. In concentrating on background, however, one runs the risk of "ghettoising" in curriculum planning. When the contextualisation of mathematical activities considers only where the students come from instead of looking at where they want to go, the notion of meaning becomes reactive (see Vithal & Skovsmose, 1997; Skovsmose, 2002b).

Meaning (for the learner) refers to a complex pattern, and to establish meaning in mathematics education presupposes that forms of negotiation are part of classroom practice. In particular, one must consider that students, as part of their dispositions, possess a meaning-producing foreground. If learners are to be understood as real-life students and not as the seemingly eager fulltime learners portrayed in several learning theories, then it is important to consider them in their "full complexity of life". This complexity provides meaning to actions, and also to actions of learning. Many studies have tried to see students' meaning construction in this broader perspective. The studies by Renuka Vithal (2003) of learning situations in a South African context and Paola Valero's (2002) studies of classrooms situations from Colombian, Danish, and South African schools illustrate what it could mean to broaden the studies of meaning in mathematics education far beyond the scope of conceptism (see also Davis, 1996; Brown, 2001).[19]

[19] It is also important to consider how many ethnomathematical studies have opened the discussion of meaning in mathematics education to include not only the meaning of concepts and notions but also the meaning of being involved in a mathematical activity. See, for instance, Powell & Frankenstein (1997). A broad conception of meaning also emerges from the discussion of adult and vocational education. An interesting discussion is found in FitzSimons (2002), even though the word *meaning* is not found in the index of the book.

To provide meaning to an educational process, it is not sufficient simply to add meaning to the mathematical concepts presented to the students. Meaning in mathematics education is not ensured even if every mathematics concept is introduced with reference to the students' pre-existing understanding. Meaning in mathematics education must be sought by involving students in meaningful learning activities. If learning is similar to action, then meaning in learning can be understood in terms of the meaning of activities. This observation implies that the notions of disposition (including foreground and background), intention, and reflection are essential, which requires planning an educational process whose purpose is made open to negotiation among the students and between the teacher and the students. Therefore, notions like framing and vantage point are relevant.

It is important that the research perspective in mathematics education is not limited in a way that would make it impossible to investigate certain relationships of meaning in the educational process. Therefore, I reject conceptism, which concentrates on meaning of mathematical concepts. This approach was prevalent when the new mathematics movement sought to ensure meaning in mathematics education. It was also prevalent when a critique of that movement suggested that, instead of bringing the students through a landscape structured by logic and set theory, they should be guided though a landscape with empirical referents for the concepts introduced. Inspired by constructivism, *meaning of concept* came to refer to what can be done by means of the concepts. According to this conceptism, essential questions to research include the following: How can students' mathematical concepts be developed further? How do the subjectively constructed concepts relate to the socially agreed-upon mathematical conceptions? Such questions, fascinating though they might be, nevertheless imply a narrow perspective on research in mathematics education.

This narrowness is disastrous because the discussion of meaning in mathematics education also relates to the politics of mathematics education, as expressed by Stieg Mellin-Olsen (1987). The students' activities, along with the meaning of such activities, have to be considered also in relation to the political world outside the classroom. If meaning is discussed according to conceptism, research will suffer from tunnel vision. For instance, in "white research" in mathematics education in apartheid South Africa, constructivism was a fashionable philosophy. That fashion was not caused by constructivism as such but rather by the accompanying conceptism, which permits educational research that is blind to the situation in which it is carried out. The discussion of meaning was separated from the political situation in South Africa, although the research was expressed in a seemingly "progressive" and "up-to-date" terminology. In this case, however, a concern for meaning in mathematics education cannot ignore that the school was one of the political instruments used to prevent black people from progressing to higher education. Conceptism is open to a corruption of the discussion of meaning. Further, an awareness of the students' foreground may include an awareness of socio-political aspects of mathematics education, an awareness that a narrow concentration on "meaning of concepts" could render impossible.

REFERENCES

Alrø, H. & Skovsmose, O. (2002). *Dialogue and learning in mathematics education: Intention, reflection, critique*. Dordrecht: Kluwer.
Austin, J. L. (1946). Other minds. Aristotelian Society *20*, 148–187. (Reprinted in Austin, 1970, pp. 76–116.)
Austin, J. L. (1962). *How to do things with words*. Oxford: Oxford University Press.
Austin, J. L. (1970). *Philosophical papers* (2nd Ed.). Oxford: Oxford University Press.
Brown, T. (2001). *Mathematics education and language: Interpreting hermeneutics and post-structuralism* (Rev. 2nd ed.). Dordrecht: Kluwer.
Christiansen, I. M. (1994). Classroom interactions in applied mathematics, courses I-II (R-94–2045a and R-94–2045b). Aalborg: Department of Mathematics and Computer Science, Aalborg University.
Cobb, P. & Bauersfeld, H. (Eds.) (1995). *The emergence of mathematical meaning: Interaction in classroom cultures*. Hillsdale (USA): Lawrence Erlbaum.
Davis, B. (1996). *Teaching mathematics: Toward a sound alternative*. New York: Garland.
Dörfler, W. (1991). Forms and means of generalization in mathematics. In A. J. Bishop, S. Mellin-Olsen & J. van Dormolen (Eds.), *Mathematical knowledge: Its growth through teaching* (pp. 63–85). Dordrecht: Kluwer.
English, L. D. (Ed.). (2002). *Handbook of international research in mathematics education*. Mahwah (USA): Lawrence Erlbaum.
Fiske, J. (1988). *Introduction to communication studies*. London: Routledge.
FitzSimons, G. E. (2002). *What counts as mathematics? Technologies of power in adult and vocational education*. Dordrecht: Kluwer..
Frege, G. (1969). Über Sinn und Bedeutung. *Funktion, Begriff, Bedeutung: Fünf logische Studien* (Günther Patzig, Ed.) (pp. 40–65). Göttingen: Vandenhoech & Ruprecht. (Originally published 1892.)
Freire, P. (1972). *Pedagogy of the oppressed*. New York: Herder and Herder.
Freire, P. (1974). *Cultural action for freedom*. London: Penguin Books.
Giddens, A. (1984). *The constitution of society*. Cambridge: Polity Press.
Habermas, J. (1984, 1987). *The theory of communicative action I-II*. London: Heinemann and Polity Press. (First German edition 1981.)
Keitel, C., Kotzmann, E. and Skovsmose, O.: (1993). Beyond the tunnel vision: Analysing the relationship between mathematics, society and technology. In C. Keitel & K. Ruthven (Eds.), *Learning from computers: Mathematics education and technology* (pp. 243–279). Berlin: Springer.
Lindenskov, L. (1993). Exploring the student's own mathematics curriculum. In J. A. Malone & P. C. S. Taylor (Eds), *Constructivist interpretations of teaching and learning mathematics* (pp. 149–156). Perth: Curtin University of Technology.
Lins, R. (2001). The production of meaning for algebra: A perspective based on a theoretical model of semantic fields. In R. Sutherland, T. Rojano, A. Bell & R. Lins (Eds.), *Perspectives on school algebra* (pp. 37–60). Dordrecht: Kluwer.
Mellin-Olsen, S. (1987). *The politics of mathematics education*. Dordrecht: Reidel.
Nielsen, L., Patronis, T. & Skovsmose, O. (1999). *Connecting corners: A Greek-Danish project in mathematics education*. Århus (Denmark): Systime.
Noss, R. & Hoyles, C. (1996). *Windows on mathematical meaning: Learning cultures and computers*. Dordrecht: Kluwer.
Powell, A. & Frankenstein, M. (Eds.). *Ethnomathematics: Challenging eurocentrism in mathematics education*. Albany: State University of New York Press.
Searle, J. (1969). *Speech acts*. Cambridge: Cambridge University Press.
Searle, J. (1971). What is a speech act. In J. Searle (Ed.), *The philosophy of language* (pp. 39–53). Oxford: Oxford University Press. (First published 1965.)
Seeger, F., Voigt, J. & Waschescio, U. (Eds.) (1998). *The culture of the mathematics classroom*. Cambridge: Cambridge University Press.
Skovsmose, O. (1994). *Towards a philosophy of critical mathematics education*. Dordrecht: Kluwer.
Skovsmose, O. (1998a). Aporism: Uncertainty about mathematics. *Zentralblatt für Didaktik der Mathematik, 98*(3), 88–94.

Skovsmose, O. (1998b). Linking mathematics education and democracy: Citizenship, mathematics archaeology, mathemacy and deliberative interaction. *Zentralblatt für Didaktik der Mathematik, 98*(6), 195–203.
Skovsmose, O. (2000). Aporism and critical mathematics education. *For the Learning of Mathematics 20*(1), 2–8.
Skovsmose, O. (2001). Landscapes of investigation. *Zentralblatt für Didaktik der Mathematik, 2001*(4), 123–132.
Skovsmose, O. (2002a). Mathematics in action: A challenge for social theorising. In E. Simmt & B. Davis (Eds.), Proceedings of the 2001 Annual Meeting of the Canadian Mathematics Education Study Group (CMESG), (pp. 3–17). Edmonton: CMESG, University of Alberta.
Skovsmose, O. (2002b). Students' foreground and the politics of learning obstacles. In the *Proceedings from Second International Congress on Ethnomathematics* (in disk format), Ouro Preto (Minas Gerais, Brazil), August 2002.
Skovsmose, O. & Nielsen, L. (1996). Critical mathematics education. In A. J.Bishop, K. Clements, C. Keitel, J. Kilpatrick & C. Laborde (Eds.), *International handbook of mathematics education* (pp. 1257–1288). Dordrecht: Kluwer.
Skovsmose, O. & Valero, P. (2001). Breaking political neutrality: The critical engagement of mathematics education with democracy. In B. Atweh, H. Forgasz & B. Nebres (Eds.), *Sociocultural research on mathematics education* (pp. 37–55). Mahwah (USA): Lawrence Erlbaum.
Skovsmose, O & Valero, P. (2002a). Mathematics education in a world apart: Where we are all together. In P. Valero & O. Skovsmose (Eds.), *Proceedings of the Third International Mathematics Education and Society Conference* (pp. 1–9). Copenhagen: Centre for Research in Learning Mathematics.
Skovsmose, O. & Valero, P. (2002b). Democratic access to powerful mathematical ideas. In L. English (Ed.), *Handbook of international research in mathematics education* (pp. 383–407). Mahwah (USA): Lawrence Erlbaum.
Valero, P. (2002). Reform, democracy, and mathematics education. Towards a socio-political frame for understanding change in the organization of secondary school mathematics. Copenhagen (Denmark), Danish University of Education, Ph.D. dissertation.
Vithal, R. (2003). *In search of a pedagogy of conflict and dialogue for mathematics education.* Dordrecht: Kluwer.
Vithal, R., Christiansen, I. M. & Skovsmose, O. (1995). Project work in university mathematics education: A Danish experience: Aalborg University. *Educational Studies in Mathematics, 29* 199–223.
Vithal, R. & Skovsmose, O. (1997). The end of innocence: A critique of 'ethnomathematics'? Educational Studies in Mathematics, *34*(2), 131–157.
Wittgenstein, L (1961). *Tractatus logico-philosophicus.* London: Routledge and Kegan Paul. (First German edition 1921.)
Wittgenstein, L. (1953). *Philosophical investigations.* Oxford: Basil Blackwell.

COLLECTIVE MEANING AND COMMON SENSE

1. WHY ARE WE TALKING ABOUT COMMON SENSE?

In a discussion of meaning in mathematics education, why should one be concerned about common sense? The answer lies in the contrast and interplay between common sense and science, where *science* is understood in the sense of a formalised system of concepts and procedures.

How can one characterise this contrast in terms of meaning, knowledge, and communication? The answer to this latter question has to start from the observation that commonsense understanding is predominantly based on correspondences, attempting to assimilate the new to the already familiar. The commonsense meaning of words depends on our ability to see resemblances or similarities. In common sense, to know means to justify conclusions already formed. In science, to explain means to establish a systematic relation between premise and conclusion.

Linguistics conceives ways of understanding or communicating as the combination of two kinds of connection, similarity and contiguity, which find their most condensed expression in metaphor and metonymy, respectively.

The polarity of science and common sense fulfils with relation to social practice exactly the same role as the different types of understanding—the analytic and the synthetic—do with respect to the individual person's cognition, or that the polarity of metaphor and metonymy does with respect to language and communication.

2. WHAT IS COMMON SENSE?

Common sense, as we conceive it, refers both to a content of taken-for-granted concepts and precepts, and to a form of reasoning that privileges the assimilation of new experience to familiar ideas.

Equally, commonsense ideas remain as prototypes rather than proceeding to definitions—permitting a conflation of strands of meaning and a permeability of conceptual boundaries. In commonsense reasoning, abduction is preferred to deduction—giving priority to conclusions over premises. Finally, in common sense, completeness takes priority over consistency—establishing the availability of a plurality of concepts and precepts so as to be capable of assimilating the diversity of experience is more important than securing their systematic compatibility. Science, in contrast, prefers consistency over completeness.

3. CAN ONE ESCAPE COMMON SENSE?

Common sense is, of course, both ubiquitous and indispensable. One cannot escape it. The foundations of any science rest on common sense, as does all human reasoning. At some point, reasoning, whether scientific or not, must rest on undefined concepts, purposes, and values that are commonly accepted. Without common sense, life as a whole can have no purpose or value. The weakness of common sense in one respect is its strength in another: That it starts from conclusions rather than premises becomes a strength when social purposes and values are at issue.

4. HOW DOES COMMUNICATION CONVEY MEANING?

Attempts to answer this question frequently stress the necessity of unambiguous definitions, of using unspecialised terminology, and of reducing the new knowledge to the already familiar. Any new information, any new knowledge, or any new idea has to be related to the corpus of knowledge already present; or in psychological terms, it must be integrable into cognitive structures one has already developed. To understand requires a perspective from which this can be accomplished. If, however, truly new knowledge is to be acquired, this perspective must be supplied, at least in part by the new subject matter itself. If something is to be introduced into thinking, this new thing must provide, to a certain degree, the standard for its own development.

As authors, writing this text, we faced the choice between working from the commonsense term *common sense* or adopting a potentially unfamiliar technical term. On the one hand, to talk of common sense would carry for the reader certain connotations and associations that broadly convey our intentions. On the other hand, because the term admits a range of possible interpretations, to use it might produce a situation in which authors and readers are at cross purposes. On balance, we have decided to adopt the familiar term because of its helpful connotations but have chosen to make explicit a more precise sense and structure.

5. HOW COMMON IS COMMON SENSE?

Not everyone shares an identical common sense in substance. One needs to understand the perceptions held by the different participants in any mathematics teaching and learning practice, as well as the forces that shape and sustain those perceptions. In any human community, whether it is as large as a country or as small as a class of mathematics students, some meanings are shared across the entire community. Often, however, the role or position of a subgroup of the community will influence their perceptions and values. The common sense of different groups may vary in complexity and sophistication. For example, both parents and teachers may have rather elaborate models of intelligence. They can label students' actions as more or less intelligent and attribute various forms of intelligence to students. The common sense of teachers regarding intelligence may be diffused to parents, and vice versa. Teachers may have an elaborated commonsense view of intelligence based on seeing groups of students of the same age behaving in different ways; parents may have an elaborated, but quite different view, drawn from their knowledge of their own children, who may differ in age and other characteristics. A teacher who is also a parent may need to develop an even more elaborate commonsense view of intelligence than one who is not, but the format or grammar of that common sense will not be inherently different.

6. WHAT IS PROFESSIONAL COMMON SENSE?

At first sight, the idea of professional common sense may appear contradictory. What we mean by this, however, is that while professionals may have a more elaborated set of ideas and experiences, the reasoning that underpins their use may retain the characteristics of commonsense reasoning as we have described it. Teacher educators are often disappointed that the scientific knowledge presented to intending teachers has only a superficial impact on their thinking. In the first instance, it can be assimilated all too easily to a common sense established during their "apprenticeship of observation" as school students. Once they embark on their classroom apprenticeship, scientific ideas are all to easily displaced by, or assimilated to, the professional common sense that they encounter. The new ideas that teachers meet as part of their professional preparation not only confront their established preconceptions but tend to be treated or manipulated within the logic of common sense, not least because of the pressures of day-to-day practice and the need to take practical action.

7. WHAT ARE THE ROLES OF COMMON SENSE AND SCIENCE IN SOCIETY?

While common sense and science are complementary, their relative social roles depend on the structure of a society. Mathematics education is part of that structure. In public (educational) discourse, the mingling of scientifically based argument and commonsense reasoning is unavoidable. Moreover, the power of the different social groups involved in this discourse is critical. The involvement of values and purposes explains why it is possible to encapsulate different perspectives without seeking to wholly resolve relatiöns between them. In order to resolve perturbations or avoid the appearance of subjective and biased political decisions, experts are called upon when the objectified underpinning of an advice from a scientific or technical sphere is need. Politicians and educational decision makers retain their commonsense conclusions, more precisely, they start with the (wanted) conclusion, and look for scientific expertise which can confirm and provide some source for it. The chapter by Keitel and Kilpatrick addresses this point.

Throughout mathematics education, in classrooms, textbooks, syllabi, teacher education courses, and treatises on didactical science, there is a tendency to ignore common sense as we have characterised it, instead of taking into account the dialectic between common sense and science. The chapter in this section recognises the need not only to acknowledge common sense but also to reshape and develop it further.

CHRISTINE KEITEL AND JEREMY KILPATRICK

MATHEMATICS EDUCATION AND COMMON SENSE

> Common sense: practical good sense gained by experience of life, not by special study. (Oxford Advanced Learner's Dictionary of Current English, 1989)

> The world of our own experience, our own reality, has split in two, and the rules applying in our daily world have no visible connection with those that apply in the realm of science. (Moscovici, 1981)

In their attempts to "make sense" learners of mathematics develop extremely specific strategies as responses to the social demands of the classroom. The strongest force of those demands is social pragmatism, and competence and meaning in the classroom are a social construction. Mathematics in the classroom has to be justified in social terms. Sense making is not just a problem of the individual learner, who needs to become able to understand, to judge, and to act. It is also a collective process for collective judgement and action. Common sense has communality, is situated, and is pragmatic. The complexity of common sense is our reason for looking more closely into the relationship between it and mathematics education.

The term *common sense* has been recently used in philosophical, linguistic, sociological, and anthropological research studies (e.g., Bourdieu, 1980; Forgas, 1981; Geertz, 1983; Moscovici, 1981), usually interpreted as a concept referring to local, situated, or everyday knowledge. The term has also been taken up in mathematics education. In his last book, Freudenthal (1991, pp. 4–9) devoted one whole section of a chapter to an analysis of the relationship between mathematics and common sense and emphasised common sense as an important starting point for mathematics learning.

Publications in mathematics education are including discussions of concepts like "everyday common sense" (Wistedt, 1994), "situated knowing" (Greeno, 1991), "cognition in practice" (Lave, 1988), or "everyday cognition" (Rogoff & Lave, 1984). We also find in the literature more studies of "children's mathematics" (Saxe & Gearhart, 1988), research on "cultural perspectives" (Harris, 1991), and comparisons of "how people think in different cultures" (Cole & Means, 1981), as well as examinations of the close relationship between mathematics and social and cultural development, including a changing common sense about mathematics in the course of history (Davis & Hersh, 1985, 1986; Joseph, 1992; Kline, 1985; Lenoir,

1979; Restivo, van Bendegem & Fischer, 1993; Solar & Lafortune, 1994). This work has raised our awareness of the importance of analysing and discussing these concepts in the broader framework of common sense.

In mathematics education, common sense is often associated with "a priori understanding", "intuition" (Fischbein, 1987; Freudenthal, 1991), or knowledge based only on subjective experience. The restriction to this epistemological aspect, in which common sense is contrasted with mathematical or scientific knowledge, is in our view misleading. It mainly serves either to reinforce a contradiction between mathematics education and common sense or to distort the substantial difference between the two.

In this chapter, we want to broaden the discussion of the relationship between mathematics and common sense by emphasising the social perspective and some distinctive aspects of this relationship. In our view, this approach can yield some insight into the problem of collective construction of meaning and sense making in mathematics education. We want to challenge some of the commonsense conceptions held by mathematics educators. These include the cult of individuality and the unquestioned universalism implicit in many conceptions of mathematics curricula. They also include assessment designs and the contradictory assumptions underlying conceptions like that of "mathematics for all". We first discuss some commonsense perceptions and some uses of the term *common sense*. Then we point out and question some commonsense beliefs held by various groups. These include beliefs by mathematicians and scientists about their disciplines, by teachers and students about the teaching and learning of mathematics and education in general, and by those who are politically responsible for the development and justification of mathematics curricula and assessments.

The major point will be to underline that neglecting the development of common sense during teaching and learning mathematics, either by implicitly or deliberately referring to it or by strongly rejecting it, hinders sense making. The neglect also contributes to a widespread aversion towards a "meaningless" mathematics and its applications. Making sense in mathematics education should enable one to develop and challenge commonsense assumptions about mathematics education. It should lead from discomfort and disconnection to comfort and connection, should relate to personal and collective experiences, and should enable investigations to become articulate. Creating and extending a new common sense in mathematics education should allow one to see meanings as multi-layered. They should be seen as collectively constructed by reflecting on what is taken for granted at all levels of the process of learning mathematics.

1. COMMON SENSE

The term *common sense* is vague. Its meaning differs across individuals as well as across cultural environments, as its equivalents in different languages reveal. Through the terms *bon sens* and *sens commun,* French offers two slightly different views of the concept. In Swedish, Dutch, and German, the terms *sunt förnuft, gezond verstand,* and *gesunder Menschenverstand*—that is, sound human reasoning—allude to its interpretation as a quality innate to the human being. The Austrian *Hausverstand* (different from the German term!) emphasizes its practical origin and destination in the home. English speakers sometimes use *horse sense* or *mother wit* in place of *common sense,* with both alternatives stressing the concept's unsophisticated, "natural" side.

Common sense is not a sense in the usual meaning of the term. Instead, it is based on evidence provided by the five senses. It begins in experience and ends in action. At the core of common sense is a process whose aim is application. Although it is often put in opposition to deeper, more demanding thinking, common sense does entail a kind of logical reasoning. The effect of this reasoning links common sense to mathematics, but its fragmentary character distinguishes the two. In contrast to mathematics, common sense takes evidence, accepted truth, and conventions as starting points and raw material for argumentation. Mathematical thinking questions its premises; common sense does not. Mathematical thinking considers forward and backward implications equally; common sense is purposeful one-way reasoning.

Nonetheless, common sense is a powerful, indispensable tool, a sine qua non for human survival. Robinson Crusoe on his desert island, for example, lacked an appropriate professional education for the necessary practical tasks confronting him. Thanks to his common sense, however, he was able to reinvent what he perceived as the essential components of his civilisation, the material and social technology of his time adapted to the island. Crusoe is a paradigm of common sense as the "natural" equipment of humans. Some philosophical and scientific research addressing this natural background of common sense relates it to an innate "operating system" of perception, understanding, and reasoning. A specialist in theories of artificial intelligence, Ernest Davis (1987), describes the indispensable aspect of common sense: "Almost every type of intelligent task—natural language processing, planning, learning, high level vision, expert-level reasoning—requires some degree of commonsense reasoning to carry out" (p. 833). Others trace it to fundamental human activities such as "counting, measuring, locating, explaining, designing, and playing" (Bishop, 1988, ch. 2). Common sense is studied as a theme of cognitive research, bearing on formal aspects of thought rather than on its subject matter aspects.

Common sense is no less crucial for survival in one's social life than on a desert island. It is the medium by which the individual maintains his or her claims against others, and at the same time mediates the balance among those claims. Common sense is the grammar of social intercourse. And it is more: The grammar is not transmitted as an abstract structure, but is fleshed out with socially accepted or current assumptions, habits, beliefs, laws and taboos, local knowledge, prejudices, and misconceptions. Thus common sense is not only a tool but also a most meaningful corpus of tradition and convention, of social or cultural knowledge and values. (Thomas Paine, 1776, referred to the corpus of tradition among the new Americans when he proclaimed his famous political goals for civil rights and a new constitution as grounded in the common sense of America's inhabitants.) That corpus is the content aspect of common sense.

Today, it is popular to see common sense as common non-sense, as old wives' tales that need to be eradicated in order to have a truly rational and scientific community. It is like the desire to purify a tribal language by correcting and enforcing the appropriate use of vocabulary and grammar. Although Descartes referred to the *bon sens* everyone had to understand his revolutionary ideas in mathematics and philosophy (Glucksmann, 1987), this trust in peoples' natural common sense has been abandoned by scientists over subsequent centuries. Voltaire's (1764) ironic reply to Descartes' assumption that "*Le bon sens est la chose la mieux partagée du monde*" ["Common sense is the best shared thing in the world"] did not share Descartes' optimism: "Common sense is not so common."

In the scholarly debates of the 17th, 18th, and 19th centuries, the idea of general education for all was proposed as a means of going beyond the conservative, restricted character of common sense. Although a political goal of the Age of Enlightenment, the idea of general education for the citizen followed an individualistic pattern. It mainly addressed the improvement of the individual's education in enlightening his or her general insight, worldview, general competence, and possible participation in popular scientific and political discourse. A contrast, therefore, between the reasoning and reflective ability of the well-educated individual scholar and the ignorant and limited common sense of uneducated laypeople served as a major argument for providing knowledge and judgment through general education.

Characterisations of common sense, especially when provided by scholars, notoriously tended to contrast it with the individual's contemplative thought. The Italian philosopher and historian Giambattista Vico (1984) saw common sense as a kind of unreflective shared judgment held by "an entire class, an entire nation, or the entire human race" (p. 63) that hindered revolutionary developments in science and politics. For Somerset Maugham (1949), it was "only another name for the thoughtlessness of the unthinking" (p. 72). A modern view in the same restrictive, contrastive vein was stated by the British author, John Berger (1967), who said that common sense was

> part of the home-made ideology of those who have been deprived of fundamental learning, of those who have been kept ignorant. [It] can never teach itself, can never advance beyond its own limits. [It] can only exist as a category insofar as it can be distinguished from the spirit of enquiry, from philosophy. (p. 102)

These characterisations reveal two important qualities today typically attributed by scholars to common sense. First, because it is "common", it is not unique to any one person. It is a social phenomenon. Second, common sense is not seen as entailing any reflective thought. Therefore it has to be distinguished from the rather different cognitive processes needed for philosophy and science.

2. MATHEMATICS AND COMMON SENSE

Historically, mathematics and common sense have the same origin: abstraction from social action based on shared sense experiences, social experiences, and social intentions. But the range and level of abstraction are different: Whereas common sense is bound to a context and meant for immediate use, mathematical abstraction as the constituent of mathematics—as an elaborated scientific theory—consistently strives to become context-free and universal. For mathematics, abstraction and normalisation are aims in themselves; structural considerations and formal rules replace content-related rules for action.

From the point of view of common sense, however, the power of abstract tools in applications has been the most convincing and effective characteristic of mathematics. Mathematics has been socially acknowledged mainly because of its potential for creating powerful instruments for multifarious uses—not only for explaining the world but also for manipulating and changing the social organisation of life and the human relationship to nature. The commonsense view of mathematics focuses primarily on its utilitarian side.

As long as mathematicians were concerned not only with abstraction and normalisation (mathematisation) but also with the reverse process of application and recontextualisation, common sense served as one means of judging and evaluating mathematical discourse. Context-bound knowledge secured one's orientation. Through the process of separating mathematics from natural sciences and human endeavours, dismissing problems of application and legitimisation as well as of social accountability in the course of specialisation, and concentrating instead on the formal refinement of prospective universal tools, mathematics and common sense became alienated from one another. They even became contradictory in their statements, and common sense was consequently seen as inferior.

Nowadays, common sense often has two quite contradictory connotations. The first is used in colloquial debates: If one would like to reject somebody's argument, one states that he or she lacks common sense. In arguing, people frequently say that their position is just plain common sense and supply no further substantiation or justification. To reject common sense then is a stroke of madness or rebellion. In science or mathematics, one can see an opposite use: Referring to common sense is not a term of praise but rather almost an insult. Many times in the course of history, mathematical developments have violated common sense; they have required that common sense be revised, mostly against strong resistance. Examples of this kind of debate include the acceptance of zero, the irrational numbers, the infinity of integers, the complex numbers, non-Euclidean geometry, and both standard and nonstandard analysis.

To underline their positive inclination toward mathematics and to distinguish mathematics from ordinary commonsense knowledge, contemporary mathematicians commonly conclude that mathematics and common sense have little or nothing to do with each other. Phil Davis (1996) argues that it is, on the one hand, obvious that

> if mathematics were simply common sense, then mathematics and its applications would be trivial. [...] On the other hand, if common sense played no role in mathematics, [mathematics] would be an incoherent concatenation of totally incomprehensible symbols. Part of the downplaying of common sense arises from the belief (misguided in my opinion) that mathematics exists in an ideal Platonic world divorced from the objects that inspire thought and from people who create and judge. Mathematics exists embedded in a [...] world of material objects and human artifacts, human language and social arrangements in which it is pursued, interpreted and validated. [...] Remove mathematics from this embedding, and hardly any piece of mathematics can survive. (p. 29)

The belief in the separation of mathematics from those who pursue, develop, and employ it; from the natural language in which it is interpreted, elaborated, and taught; and from its primitive conceptions, strategies, philosophies, and applications has led to a narrowing of common sense. Common sense is seen as a common acceptance or justification of foundations as well as of convention within mathematics. There is a specialised common sense among practising mathematicians that it is crucial to ensure a common sense among outsiders concerning the success of the whole enterprise of mathematics.

2.1 Consensus versus reification

Our realities, whether physical or social, can be partitioned into what Moscovici (1981, p. 186) terms the *consensual* and the *reified* universes. Our consensual universes are realms in which we experience our mutual humanity. Society's own goals and meanings predominate; we create a "community of meanings" through discourse and social interaction. Our knowledge is judged against our collective purposes and values. Everyone participates on an equal footing. Moscovici uses the term *social representations* to characterise these consensual universes. *Social representations* are the

> concepts, statements and explanations originating in daily life in the course of inter-individual communications. They are the equivalent, in our society, of the myths and belief systems in traditional societies; they might even be said to be the contemporary version of common sense. (Moscovici, 1981, p. 181)

Our reified universes, in contrast, are those domains of practice in which we attempt to submerge our individual identities and to distance our works from human frailties. We seek fundamental and immutable knowledge, judged according to authoritative rules for determining validity. Some of us, consequently, become more qualified than others to make these judgements. As Moscovici says, "Obviously, science is the mode of knowledge corresponding to the reified universes and social representations—common sense—the one corresponding to the consensual universes" (p. 187). In other words, mathematics and common sense, in this analysis, belong to separate spheres of reality.

Common sense makes use of two processes of colloquial reasoning: *anchoring* and *objectification* (Moscovici, 1981, p. 192). Learners attempt to make the unfamiliar familiar in commonsense ways. Either they attempt to anchor it to what they already know, fitting it into their existing schemes for categorising phenomena and thereby confirming their preconceptions, or they objectify it so that it becomes almost physical and thereby under their control. In school, the regular patterns one sees in classroom activity stem in part from the shared systems of constructs held by the participants in the process. For example, one of the pupils' primary tasks in school is to use their common sense to learn how these systems operate and to internalise them.

Another example of anchoring and objectification has been described by Freudenthal (1975): When teachers were asked to rate test questions from the First International Mathematics Study according to how much opportunity their students had had to learn the content, they anchored the unfamiliar "opportunity-to-learn construct" in the familiar idea of difficulty and ended up rating questions according to their estimates of how many of their students would answer them correctly. In much the same fashion, users of schemes for classifying test questions by cognitive level (such as that of Bloom, 1956) often find themselves using estimated difficulty as the anchor. In an interesting illustration of how ideas flow back and forth between the consensual and reified universes, constructs such as *opportunity to learn* and *cognitive level,* having become reified in educational research, have been taken back into educational discourse as elements of reality. Similarly, intelligence, through the measurement of the "intelligence quotient", has become objectified in commonsense understanding (Ruthven, 1996).

To improve common sense about mathematics education requires attention to the consensual and reified universes and their interplay. The mathematician's specialised knowledge of mathematics needs to be connected to the public discourse about mathematics. The popular views of mathematics teaching pervading our culture need to inform and to be informed by the educator's professional understanding. We need to refashion our social representations so that we can see mathematics education in a different light:

> The challenge for mathematics education [...] is to reconstruct common sense through dialogue between the consensual and reified universes; to provoke change not only in the constructs and principles guiding the everyday thinking of teachers, pupils, parents and politicians, but in the systems of curriculum and pedagogy which frame that thinking. (Ruthven, 1996, p. 111)

2.2 *Application and evaluation of mathematics*

It is a commonplace that the devices of mathematics, as well as the technology it spawns for shaping and formatting our world, are beyond the reach of commonsense judgement. The field is left to professionals and specialists in their respective domains. What about accountability? Who is to be held accountable?

For centuries, one of the great values of common sense was that its reasoning served as a basis for public discourse on matters of common interest. That discourse enabled people to evaluate and control the means by which knowledge was being applied (see Habermas, 1971, for an examination of public discourse and rational analysis in a democratic society). With the abdication of common sense as an authority in the sophisticated processes of organising modern societies, the commonweal has lost its most ubiquitous instrument of communication and judgment.

That loss effects explicit mathematical applications and, even more, implicit mathematics. (The term *implicit mathematics* refers to the transformation of mathematical concepts, theories, and models into social or material technologies; into "social algorithms" such as military regulations, working instructions, and legal statutes; into institutions of social practice like bureaucracies; and into all kinds of autonomous machines and information technologies; see Keitel, Kotzmann & Skovsmose, 1993; and also Weizenbaum, 1976, who calls for a new common sense toward mathematical technology.) Implicit mathematics penetrates social life in all domains and at all levels. Its influence largely passes unnoticed, and in particular its underlying aims and purposes remain concealed. The defeat of common sense leaves the design and installation of mathematical devices in social life uncontrolled and unevaluated. Common sense follows its very nature: It takes the pragmatic path.

Solutions to the problem of control cannot be expected to come from society's political representatives. To achieve their objectives, politicians readily and uncritically adopt specialists' offers of social and material technologies. Insofar as politics is directly confronted by the role of mathematics in the modern world, the politician's reaction is one of helplessness. Thus, it is often argued (Australian Education Council, 1990; Cockcroft, 1982; National Council of Teachers of Mathematics, 1989) that for the sake of technological development and scientific progress—which is unthinkable without the ongoing mathematisation of our society—more mathematical knowledge ought to be acquired by more people, not only for their vocational and professional success but also for their social functioning. In reality, however, the transformation of mathematics and other expert knowledge into technology fosters a kind of mathematical disqualification: The more implicit mathematics there is in society, the less the explicit doing of the mathematics provided in the mathematics classroom is either required or practised in daily or professional life.

2.3 Instrumental knowledge versus orienting knowledge

Mathematical activity draws upon a wide spectrum of competencies. Does society provide for the right ones? And does society simultaneously provide, on the one hand, for both the renewal and extension of expert knowledge of mathematics and, on the other hand, for the development of ordinary commonsense knowledge of mathematics? The German philosopher Mittelstrass (1982) proposes a distinction between two types of knowledge. The first is an instrumental knowledge

(*Verfügungswissen*), that is, a knowledge that is at one's disposal. He describes it as the mastering of techniques and skills that can be acquired by dealing with concepts in pure mathematics as well as with methods of applying those concepts (a mapping and modelling knowledge). The second is a directing or orienting knowledge (*Orientierungswissen*) that is related to metacognition and hermeneutic methods. Mittelstrass characterises it as heterogeneous, evaluative, normative, and justifying. It searches for overviews, connections, and meaning; in particular, it aims at a critique of ideologies and their negative effects. Where and by what processes do we get orienting knowledge when the improvement or development of common sense in education is neglected and common sense is kept as something static, immobile, and therefore outdated by the rapid changes of modern times?

Clearly, common sense is instead something established anew by each generation, with its own starting point, perspectives, and challenges. The technology available in society plays a major part in this renewal. For operations covered by instrumental skills, common sense draws on technology. Whereas a century ago, for example, paper-and-pencil algorithms provided the medium of choice for performing arithmetic computations, anyone today with common sense who needs to perform such operations will make use of a hand calculator. Today's common sense makes use of today's technology, more rapidly even than today's scientific thinking or education. Therefore, at least in certain aspects, common sense is neither static nor revolutionary but rather changes as a function of societal change. But do we address common sense in our teaching?

Common sense would be better if it were supplied with orienting knowledge and not with technical knowledge only. If education is to contribute to the transformation of common sense, it needs to develop orienting knowledge. Curiously, many efforts in mathematics education neglect orienting knowledge altogether, assuming that technical knowledge will suffice. The fundamental question seems to be: How do we educate common sense? How might mathematics education make greater efforts to strengthen it?

3. TEACHING AND LEARNING MATHEMATICS

The relationship between school education and common sense is more complicated than it might first appear. Common sense infiltrates the actions of all participants in education. It affects the goals and the means of instruction (Gellert, Jablonka & Keitel, 2001). It influences the content taught and the methods of teaching it at all levels (Clarke, 2002; Clarke et al., 2002; Keitel, 2002; Shimizu, 2002; see also http://www.edfac.unimelb.edu.au/DSME/lps/). When children enter school, they are already full of common sense. Education then begins both to build on their common sense and to replace it. Education in all disciplines—and mathematics education in particular, since it relies on anticipatory reasoning—depends strongly on common sense. Moreover, the ultimate outcome of schooling, insofar as it is not concerned with the knowledge needed by experts (as in the upper secondary grades), is the attainment of a new standard of "sound" common sense.

Mathematics education is a social enterprise. It is concerned with socially acknowledged and selected knowledge for socially determined aims and goals. The knowledge selected to be taught depends on common sense within the community of mathematicians, mathematics educators, and those responsible for the educational system. The classroom as a major part of the didactic system creates relationships between individual and group patterns of communication, as well as between individual and social knowledge and meaning in mathematics. In short, school education, including mathematics education, is the most effective agent in the reproduction of common sense, regardless of whether this role is recognised or accepted. It may not be accidental that a crisis of common sense regarding the knowledge to be taught in school has coincided with a decline in the concept of general education.

3.1 Meaning and communication in mathematics education

Communication has been a prerequisite to the development of mathematics. Mathematics is constituted through the communication of ideas. Over the centuries, mathematical and metamathematical discourse guided by explicit and (finally) implicit rules has determined a code that conveys unmistakable meaning in mathematics. In mathematics education, the learning process goes on at two levels, each with its own language. There is the language of everyday life and the more formal language of academic mathematics. The colloquial commonsense language serves as the substratum on which specific mathematical communication gradually grows.

Mathematical language is not the only language that grows. Colloquial language grows, too, and the building of proficiency indeed starts from very provisional structures. By consciously confronting colloquial language with mathematical language, mathematics education offers a rare chance for teachers to address common sense directly, along with its implications and limitations, even with young pupils. In the lower grades, discussions of zero, fractions and division, operations with negative numbers, and relationships between area and length may offer favourable opportunities for teachers to promote pupils' awareness of common sense and language. The transfer of fragments of discourse from the level of professional language to the level of colloquial language, which is frequently used in political rhetoric to intimidate and to overwhelm one's opponents, may be addressed in the higher grades to make pupils aware of the problems of orienting knowledge.

One type of communication occurs in the classroom when teachers attempt to assess what students have learned, either formally or informally. Any assessment activity involves communication between teacher and students about what is expected; what sorts of response constitute superior, acceptable, or unacceptable performance; and what the consequences are likely to be. Assessments tell students what mathematics learning is valued. They show the mathematics that the teacher or others think is important for the students to know and remember. They give students information and judgements about mathematics itself and about the students themselves as mathematics learners. One of the six "standards" for assessment advanced by the National Council of Teachers of Mathematics (NCTM, 1995) says, "Assessment should enhance mathematics learning" (p. 13):

> Assessment is a communication process in which assessors—whether students themselves, teachers, or others—learn something about what students know and can do and in which students learn something about what assessors value. When the focus and form of assessment are different from that of instruction, assessment subverts students' learning by sending them conflicting messages about what mathematics is valued. When instruction pursues one set of goals and the assessment—especially if it is for high stakes—pursues another, students are faced with a dilemma and must assume that the goals of assessment are the ones that count. (p. 13)

Mathematics assessment often works against good instruction because the assessment process sends a subversive message to students. Teachers may proclaim the importance of carefulness and then accept and reward careless work on projects. They may stress the importance of deductive reasoning in mathematics and then give examinations that require little more than rote memorisation. Students learn to "read" what counts in an assessment situation and to disregard what the teacher might be saying about what they need to learn. They acquire their own common sense about how assessment will proceed in the school context. They have internalised not only the overt messages the school is sending but also the covert ones.

3.2 Common sense and teaching mathematics

A Dutch study (van den Heuvel-Panhuizen, 1994) showed that many preschool children can successfully solve problems that belong to the curriculum of Grade 1, although these problems had been judged by mathematics educators and experienced teachers as impossible for children of this age to solve. Unspoiled by standard algorithms, the young children used, of course, only informal strategies based on common sense.

During childhood, ideas and conceptions of the world evolve as a result of children's experience and socialisation into commonsense views. When children enter preschool or primary school, these views serve as a means for sense making and the construction of meaning. At the same time, primary teachers make a pragmatic appeal to common sense as a basis for explanation and justification in facilitating the learning of mathematics. As long as mathematical operations or concepts are closely related to concrete action and contexts, commonsense arguments referring to the use of one's "common senses" ("You see what you have done") seem to be the most helpful means for promoting understanding. Mathematics can then be explained by teacher and pupils as the generalisation of different informal strategies that are compared, discussed, and confined to a "standard." There is a common belief among future primary teachers that a major part of what is traditionally taught and still is the dominant content of the primary curriculum in mathematics can be seen as mere common sense.

But commonsense ways of explaining phenomena or judging ideas and actions always refer to uses for specific purposes or in specific situations, and conflicts with commonsense arguments—already in primary school—originate in learning situations when concrete actions and contexts are left behind and have to be transformed into mathematical generalisations, formalisations, and abstractions. The contextualised understanding of arithmetical operations becomes a cognitive obstacle, and the plea "Use your common sense and you can see it" then becomes irrelevant. Commonsense experiences begin to counteract understanding and meaning construction. The pupils' "lack of common sense" is then blamed by the teacher for their misunderstanding, as is their inappropriate use of common sense: "Mathematics has nothing to do with common sense."

The terminology used in mathematics and mathematics education represents a particular example of misunderstanding in reference to common sense. The term *natural number* has its connotations in colloquial language, and most children accept it when encountering systematically the natural numbers in the classroom. Today we know that our natural numbers are by no means "natural", but developed according to certain demands or conditions of the social constitution of a community or society, depending on concepts like hierarchy, private property, and the complexity and differentiation of the social organisation of labour. A certain level of social organisation required processes of quantification. Then counting, measuring, and calculating became important, necessary, and used. Studies of tribal groups show that the invention of numbers was not necessarily connected to the development of number systems: Some groups had numbers and counting procedures, but the use of numbers was restricted and numbers were not systematically connected. They did not play the same important role in social organisation. Instead, habits and magic or religious rites served that function.

Nissen, Damerow, and Englund (1990), who are studying and decoding Uruk tablets—the first texts containing numbers and number systems with operations for mapping and solving general problems of administering the economy in a complex and highly hierarchic society practising slavery—convincingly describe how the professional need to deal with numbers required a systematic approach and how the use of the number system became a kind of common sense for the dominant social groups of scribes. But common sense for a few can still be magic and uncommon for the majority. During the Renaissance, banking systems demanded complex operations with "negative" numbers ("red" versus "black" numbers in double-entry bookkeeping), which by a systematic formalisation and mathematisation could be described much later as extending the system of positive rational numbers by adding the negatives, an extension that struggled a long time for general acceptance. By the transfer of the imaginary system of money from banking systems to spheres of production (applying double-entry bookkeeping to the production sphere became standard during the Industrial Revolution), practical dealings with "negative numbers" ("costs") became accepted as part of a new common sense, along with philosophical debates about the "strange entities and rules": The practical use took priority over the theoretical and philosophical foundation.

A major obstacle arises when pupils encounter zero as nothingness or emptiness resulting from a concrete action, but need simultaneously to accept zero as a real number and a cipher of great importance for the decimal place-value system and for calculation. Although "nothing", this number has to fulfil specific rules if it is to be used in both aspects. Zero is unavoidable and provides strange results. As long as addition or subtraction is involved, this use does not harm pupils very much. Because zero is the neutral element of addition, it can be related to materialised actions. But multiplication and division cannot be explained by using the frame of emptiness, as this goes far beyond a commonsense reference to concrete action.

Ambiguity in references to common sense (both calling upon it and rejecting it or raising obstacles to it) is frequently experienced by students in school. Studies of pupils' errors (e.g., Baruk, 1985) and their anxiety (e.g., Buxton, 1984) refer to this ambiguity when explaining increases in aversion toward mathematics and in mathematical errors. The ambiguity helps create negative attitudes—in particular, in future primary teachers—that are difficult to either challenge or alter. It contributes not only to math anxiety and low performance, but also to a general mistrust of mathematics, which is experienced as a meaningless and incomprehensible subject.

Mathematics teaching has too seldom encouraged a broad and lively development of children's common sense along with a study of the discipline—even though mathematics, with its impact on practically every area of modern life, can offer particularly rich opportunities for such a development. Instead, in primary school mathematics, the handling of the basic operations of arithmetic is still the dominant activity. The usual exercises, as well as the so-called "applications" in word problems, reduce mathematics to a "numbers game"—find the operation and execute it! Calculating perimeter or area becomes the major concern in geometry. In secondary mathematics, equations, functions, or calculus problems are rarely analysed theoretically. Constructions similar to "word problems" are even found in higher-level courses, and for many university students linear algebra is just calculating with matrices. Instructional practice of this type, although resembling a sort of commonsense approach because of its modest expectations for student learning, is in fact no more than convention and boring routine. Pupils' common sense is most likely to develop in opposition to such practice.

On the other hand, the hidden curriculum of teaching and learning in schools also creates a common sense for pupils that mostly passes unnoticed by teachers. After a few weeks, a classroom rubric develops in which pupils and the teacher have learned to read each other's signals, to know when things are serious or light-hearted. A mathematical atmosphere is then established that frames all activities and communication. What is uncommon to the pupils but common for the teacher becomes common sense taken as common, as progress in learning. Pupils and teachers are locked in a contract in which "the more clearly the teacher indicates the behaviour which would follow from understanding, the more easily the pupils can display the behaviour without generating understanding" (Mason, 1996, p. 155). It is always amusing to witness how children develop common sense in their professional life as pupils. They learn very early how to organise their work. They make the broadest possible interpretation of restrictions and formally meet the teachers' expectations while making their work as economical as possible. They also learn to use all kind of technical equipment. They become psychologists who can anticipate what their teachers need to hear and what they cannot stand. Children find food for developing common sense everywhere, but the result may be different depending on whether the school is able to take up the process and carry it far beyond the children's limited horizon, or whether children develop it in opposition to the school, which may be a very unfortunate experience.

3.3 Common sense and curriculum development

Mathematicians view the mathematical object to be taught or learned as structurally but not qualitatively the same as that object within mathematics. After the "logical" or "reasonable" selection of mathematical objects, the "simplification" or "elementarisation" of content as a process of transforming mathematical objects into objects of teaching and learning is regarded as the responsibility of mathematics educators. What is reasonable or justifiable refers primarily to a common sense—to conventions or agreements within mathematics or to common sense arguments outside of mathematics. Most mathematicians believe that mathematics education is simply concerned with problems of the type "How do we tell the important mathematical facts to children?" It is not surprising, therefore, that publications appear periodically to great public acclaim that try "to fill the gap" by offering the ultimate in a basic curriculum for school mathematics. Such curricula invariably contain only those portions of mathematics that are already part of common sense (Slavin, 1989).

This process may be explained by the concept of *didactic transposition* (Chevallard, 1985), which aims at analysing the "basic laws" of this transformation of knowledge into curricula and programs for teaching. The underlying common sense of this concept is the assumption that starting within mathematics (the mathematics of professional mathematicians at the university) is the natural way to obtain any reasonable body of school mathematics. The transposition is understood as the whole process of selecting, analysing, re-interpreting, and changing objects taken from mathematics into objects of teaching and learning as types of knowledge to be taught. Although this approach is thought to apply to the whole of school mathematics, its origin in secondary school mathematics is obvious. Related approaches indeed have strong traditions and support in the didactics of secondary school mathematics in, for example, Germany and France. The social context there has a strong culturally based consensus that mathematical activity is good as such and that it must be supported even if it seems on its surface to be useless. On the level of justification of a curricular design, the goals and arguments may then change substantially, but the changes in content are automatically interpreted in the frame of this conservative common sense about mathematics education. The social use of (implicit and explicit) mathematics is not considered. It does not influence curricular decisions beyond a superficial level of political justification.

The social use of mathematics has little or no status in usual didactic approaches, and the didactic transposition does not reflect it at all—not through inadvertence, but in line with the underlying commonsense philosophy. That philosophy rests on (a) an academic presumption as to the absolute value of disciplinary knowledge over social knowledge, (b) a conviction that a thorough general grounding in disciplinary knowledge best covers all the needs and applications the pupil will encounter in the real world, and (c) an interest in the early cultivation of professional thinking in and positive attitudes toward the discipline of mathematics (even though only a tiny minority of pupils will ever become professional mathematicians). The resulting curricula are often neither "logical" nor "reasonable": Whether curriculum content is organised into a hierarchy of concepts, an accumulation of topics, or a clustering of problems, it can extract only limited selections out of the "quarry" of mathematics. The image of mathematics in the curriculum is thus partial and consequently easily distorted.

The curriculum follows unquestioned convention and the principles of a post hoc logic within the discipline rather than any epistemological or psychological order of learning processes. Goals referring to social needs and demands are disconnected from the content. Questions of "why mathematics education" and "what mathematics for all" are answered only with common sense about the universal use of mathematics being taken for granted, which is neither substantiated nor can be experienced in the classroom or outside, whether in daily life or in one's professional life. Here the common sense of the professionals hinders the development of a new design of "mathematics for all"; in particular, if it is meant for nonindustrialised countries or societies that are trying to establish a common education system.

3.4 The common sense of assessment

The assessment practices mandated by political authorities provide a particularly striking illustration of how commonsense views of instruction play out in school practice. What could be more commonsensical than the notion that schools, like businesses, need to be held accountable for the quality of their "product"? From this platform of accountability—with the underlying metaphor of school as a "factory" that "produces" outcomes—one easily moves to the view that instruction is a kind of manufacturing process in which students acquire curricular objectives step by step. Each step can be identified and its attainment assessed. Once a framework of objectives has been laid out, accountability efforts can focus on which students have attained which objectives. Students, and schools, can then be ranked, and rewards (or punishments) distributed.

A major difficulty with this approach is that the stepwise model of educational attainment does not fit what we know about how students learn (Hoyles & Sutherland, 1989; Resnick, 1987). Students do not learn mathematics step by step, moving up a hierarchy of concept difficulty or cognitive complexity. Their learning draws instead a melange of memorising, reasoning, and problem-solving skills as they develop their own strategies and impose their own organisation on whatever curriculum has been laid out for them. Ruthven (1996) has shown how efforts at national assessment in England moved from a conception of progression through objectives arranged in a hierarchy to a conception built on progression by simple accumulation. The attempt to impose a hierarchy did not work: Students failed objectives at lower levels than others they had passed, and their patterns of performance showed other inconsistencies. The assessment authorities were obliged to take a more global, less analytic view of progression.

In a parallel fashion, the view of measurement embodied in the English national assessment shifted from denotation (actual performance on individual objectives) to connotation (idealised performance on complex amalgams of objectives). Authorities no longer claimed that pupils' achievement could be linked to specific objectives. Instead, pupils were assigned to "levels" according to a kind of best-fit procedure that suppressed anomalous evidence (Ruthven, 1996). The assessment process was thus rendered "more credible", in the words of one of the developers. Ruthven concludes, "here we see both the resilience and the plasticity of common sense" (p. 109).

4. COGNITIVE AND EPISTEMOLOGICAL RESEARCH RELATED TO COMMON SENSE

Common sense and its relation to mathematics education have not been an explicit subject of research, but they should nevertheless be addressed directly. A first step would be to summarise related approaches. There are research directions, in particular in neighbouring studies of cognition and epistemology, that may cast light on the subject.

Within the development of artificial intelligence, for the construction of intelligent tutorial systems or interactive learning programs, common sense has become an object of research.

> How to endow a computer program with common sense has been recognized as one of the central problem of artificial intelligence since the inception of the field. [...] Common sense involves [such concepts as] quantity, space, time, physics, goals, plans, needs, and communication. [...] To design [an intelligent] system, we must create theories of all the commonsense domains involved, determine what kind of knowledge in each domain is likely to be useful, determine how it can be effectively used, and find a way of implementing all this in a working computer program. (Davis, 1990, pp. 1–2)

Common sense viewed as a kind of natural property of the human being, the "operating system of the brain", is simulated and analysed in order to provide insight and to develop new formalisations of human intelligence and knowledge. Research in cognition allows for both new tools and new subject areas.

A variety of mathematics education research studies have emerged from the notion that there is an unbridgeable discrepancy between what certain social groups or countries need and what established mathematics education has to offer. Starting from the basic assumption that learning and understanding are essentially determined by local, cultural, and social influences, several research perspectives for mathematics education have developed that differ more in focus than in direction. These include perspectives based on ethnomathematics (e.g., Harris, 1991; Zaslavsky, 1987), feminist theory (e.g., Belenky, Clinchy, Goldberger & Tarule, 1986; Solar & Lafortune 1994; Walkerdine, 1988; see also Pimm, 1991), constructivism, and socio-political approaches operating with concepts like "local knowledge", "everyday understanding", "situated cognition", or "the social construction of meaning". The locally, culturally, and socially based intelligence that is observed in social practice, and that is independent of school learning, can be identified as common sense. The case studies provided by researchers adopting these perspectives may add considerably to our understanding of common sense and its relation to mathematics education.

4.1 Conflict and cooperation between common sense and mathematics education: The concept of zero

In a study with Italian students, A. Codetta Raiteri and E. Caianello (1996; see also Capucci, Codetta Raiteri & Cazzaniga, 2001) tried to link commonsense views to cognitive and emotional dimensions of mathematical activities. Although they did not aim at statistically based findings, the research covered a large number of Italian regions and involved 2500 students at all levels from primary to university, as well as the collaboration of hundreds of teachers. Starting from the assumption that all students build up opinions, concepts, naive theories, meanings, and explanations about the various aspects of their school experience and that these form the bases for interpreting, receiving and organising new knowledge, the researchers designed a questionnaire with the same questions for students from each level of school and for some adults. These questions concerned the relationship between crucial mathematical concepts and commonsense notions in different age groups. The questions were to provoke not only cognitive responses but also emotions and some practical ideas through a variety of response forms (short and extended text, designs and drawings, etc.). In the first part of the study, the questions were *What does zero mean to you? What difference would it make to you if there were no zero?* The questions were put in different contexts to create a semantic net and to evoke ideas and images linked to zero.

The researchers justified their choice of questions about zero by noting the very special role it has played in history. The invention and—after strong resistance—acceptance of zero in the social history of mathematics is seen today as the greatest revolution in the development of the concept of numbers. Zero revolutionised human thinking and common sense as well as social constitutions and developments. It led to imaginary money in the mercantile banking system and gave rise to precursors of negative numbers in the economic practice of production control. A long road leads from the antagonism between early Christian culture and the symbols of "nothingness" or "emptiness" used by the Greeks (the "*ex nihilo nihil fuit*" of Genesis versus the "*horror vacui*") to today's taken-for-granted operations with zero as a fully accepted number and an important letter in the most widely used place-value system. This change is connected with the necessary abundance of esoteric symbols associated with the sign of zero (the hollow crown of the Cabal, the circle of light as infinity). To summarise Codetta Raiteri and Caianello's (1996) preliminary conclusions:

- The great uncertainty students show with respect to the meaning of zero and its possible uses is mainly connected to such common views as 'Zero is nothing', 'Zero connotes emptiness', or 'It is the absence of any thing or symbol'.
- In many students and adults, zero stimulates strong emotional responses related to a fear of emptiness or of the unknown, while it evokes in others thoughts of equilibrium and growth.
- The emotional associations and meanings of zero lead students to build up complex cognitive nets around the concept that are only rarely acknowledged or supported by school learning.
- Mathematically 'sound' or justifiable meanings of zero are dominant in students' thinking only when they are explicitly dealt with in the classroom; however, they are partially covered by layers of naive but strong commonsense views and negative feelings.

Codetta Raiteri and Caianello propose that, in particular, in the teaching and learning of zero, emotions have to be addressed explicitly. The teacher has to pay attention not only to the map of mathematical concepts but also to those cognitive-emotional nets that students build up spontaneously by connecting them to unexamined commonsense views and feelings.

A follow-up study at the Free University of Berlin (Jablonka & Gellert, 1996) explored a similar research question about zero within the restricted domain of pre-service teachers' beliefs and commonsense views. Pre-service teachers were asked to write essays about the two questions and to reason about their answers. Unlike the students in the Italian study, these prospective teachers often had to refer back to buried experiences and times long before when they had learned about and operated with zero in the classroom. What was striking were the very similar results.

These future primary school teachers were usually not able to design a mathematical map of zero. They exhibited many emotional expressions of anxiety and fear of the "nothingness" and "emptiness" that they associated with zero. These associations prevented them from explaining how to operate with the concept of zero in mathematics in general, and sometimes even not in arithmetic. Beyond some cautionary rules ("Division by zero is strictly forbidden"), all of their explanations referred to concrete actions in everyday experience that proved to be impossible, revealing major misunderstandings. The complaining sigh "If it is easy, it is common sense; if it is difficult, it is mathematics!" could be heard as they expressed their distance from mathematics.

The essays provided a good opportunity for them to question common sense, to take commonsense views as a starting point for epistemological inquiry and the generation of new wisdom. The prospective teachers anticipated that common sense can be probed and challenged to reveal aspects that are being stressed or ignored, to create and attempt to resolve tensions, and to raise their awareness of the foundation of their own thinking and arguing. This process not only has to happen with mathematics, but is also fruitful with mathematics education for teachers. In their discourse with the prospective teachers, the researchers went further and applied this reflection process to typical commonsense statements from education such as:

- Always start where the pupils are.
- Build up pupils' knowledge from the simple to the complex.
- Base mathematics on real contexts (see also Mason & Monteiro, 1996, pp. 98–100).

The researchers found many counter-arguments and counterexamples to these commonplaces that could be used as eye-opening activities in different instructional contexts.

5. THE IMPACT OF SOCIAL CHANGES

General schooling has always been oriented towards the future working lives of pupils. Complaints about the success of the preparation given them have been numerous, but the goal has never been questioned. What does it mean for schools that successful learning no longer leads more or less automatically to vocational or professional careers? Or that careers eventually may occupy but a relatively short part of one's life? Can, or should, general schooling maintain its dominant orientation towards work? Is the balance between knowledge and competence related to common sense affected by these changes? Is disciplinary knowledge affected? What are the consequences if life outside employment—itself increasingly mathematically structured—is to make sense, when mobility has become crucial and when a repeated change in professions during one's working life has become the norm? Comparative studies, such as those conducted by the Organisation for Economic Cooperation and Development (Black & Atkin, 1996), describe general tendencies in all industrialised countries towards "redefining subject empires" and strong movements towards new approaches of designing "mathematics and science for all."

The lack of labour in Western industrial societies is one problem affecting the goals of schooling. Compared with the tremendous upheavals in other parts of the world, however, it is a problem of relatively limited dimensions. Worldwide social changes and the transformation of political systems—for example, in Middle and Eastern Europe or in South Africa—entail changes and new demands for education;

in this process, mathematics and science as school subjects usually play an important role. In some countries, many hopes have been placed on mathematics and science education for the development of the new society—in South Africa, the call is for "mathematics and science for all and for a democratic society." In other countries, such as those in Eastern Europe, social changes have called into question the previously privileged role of mathematics and science education in the curriculum. Or at least a change in the focus of these subjects is demanded. Mathematics educators in several countries are joining in efforts to develop approaches of "critical mathematical education" (Skovsmose, 1994; see also Frankenstein, 1989) as a basis for democratic competence. Movements for equity in mathematics education address in particular the need to overcome racial, gender, and social-class barriers to learning.

Efforts to promote mathematics for all (see Damerow, Dunkley, Nebres & Werry, 1984; and in particular, Damerow & Westbury, 1984) entail a reconstruction of the content and methods of mathematics instruction so that common sense can play a more prominent role. Mathematics as reasoning (Walkerdine, 1988) is extended to include mathematics as social practice. Pedagogy recognises that cognitive change requires management and modification of the social environment (Newman, Griffin & Cole, 1989). The school mathematics curriculum becomes a site where common sense is valued, used, and developed more fully.

6. CONCLUSION

In the discourse of mathematics education, the topic of common sense is seldom addressed directly, although there are many indirect references, often in ways that set mathematics above or beyond common sense. Few articles or books explore the common sense of mathematics teaching and learning. Few research studies address commonsense knowledge of mathematics, what it is, and how it might be changed. This chapter has attempted to put common sense on the agenda of mathematics education.

If common sense is to play a greater and more appropriate part in mathematics education, it needs to be seen more clearly for what it is and what it is not. One of its strongest roles can be to offset the specialisation at which so much mathematics instruction aims. Common sense is not a level of attainment that one reaches and then transcends. Instead, it stands apart from the school disciplines and from codified knowledge. Advanced studies in mathematics need to build on and to develop common sense, as far as possible. They should not attempt to replace it.

The vision of "mathematics for all" demands that school mathematics be enriched and questioned by common sense to a much greater extent than one typically finds today. Education for a democratic citizenry is enhanced when mathematics and common sense combine to become a forceful instrument for orienting pupils' knowledge.

Finally, common sense is sometimes quick and insightful. It cuts below surface qualities. It is playful, forging clever connections between otherwise incongruous ideas. In a word, it has *wit* and could be used to sharpen one's wit. Consequently, it provides a welcome complement to much of the mathematics institutionalised in school.

- Common sense is not a customary concept in mathematics education or in discussions of research; this chapter asks for a change in that situation.
- Common sense seen as an enrichment of school mathematics refers to the vision of "mathematics for all."
- Common sense is not a level of attainment; it stands apart from the disciplines and is not to be replaced by higher studies in mathematics.
- Common sense provides a counterbalance to specialisation.
- Common sense and mathematics education could combine to forge a powerful instrument for orienting the knowledge needed by a democratic citizenry.
- Common sense has wit and thereby complements much of school mathematics.

REFERENCES

Australian Education Council. (1990). *A national statement on mathematics for Australian schools.* Carlton (Australia): Curriculum Corporation.

Baruk, S. (1985). *L'age du capitaine: De l'erreur en mathématiques* [The age of the captain: On error in mathematics]. Paris: Editions du Seuil.

Belenky, M. F., Clinchy, B. M., Goldberger, N. R. & Tarule, J. M. (Eds.) (1986). *Women's ways of knowing: The development of self, voice, and mind.* New York: Basic Books.

Berger, J. (1967). *A fortunate man: The story of a country doctor.* Hammondsworth (England): Penguin.

Bishop, A. J. (1988). *Mathematical enculturation: A cultural perspective on mathematics education.* Dordrecht: Kluwer.

Black, P. & Atkin, J. M. (Eds.). (1996). *Changing the subject: Innovations in science, mathematics and technology education.* London: Routledge.

Bloom, B. S. (Ed.). (1956). *Taxonomy of educational objectives: The classification of educational goals. Handbook 1: Cognitive domain.* New York: McKay.

Bourdieu, P. (1980). *Le sens pratique* [Practical sense]. Paris: Editions de Minuit.

Buxton, L. (1984). *Do you panic about maths? Coping with maths anxiety.* London: Heinemann Educational.

Capucci, G., Codetta Raiteri, A. & Cazzaniga, G. (2001). *Lo zero e il senso commune: Rapporto di ricerca sullaprovvisorietà di un apprendimento disciplinare* [Zero and common sense: Report of research on the transience of disciplinary learning]. Rome: Armando Editore.

Chevallard, Y. (1985). *La transposition didactique: Du savoir savant au savoir enseigné* [The didactic transposition: From scholarly knowledge to knowledge taught]. Grenoble: La Pensée Sauvage.

Cockcroft, W. H. (Ed.). (1982). *Mathematics counts.* London: Her Majesty's Stationery Office.

Codetta Raiteri, A. & Caianiello, E. (1996). Does mathematics teaching influence common sense? Does common sense interfere in mathematics learning? In C. Keitel, U. Gellert, E. Jablonka & M. Müller (Eds.), *Mathematics (education) and common sense: The challenge of social change and technological development* (Proceedings of CIEAEM 47, pp. 143–151). Berlin: Free University of Berlin.

Clarke, D. (2002). *The learner's perspective study: Exploiting the potential for complementary analyses.* Paper presented in the symposium "Primary Research, Secondary Research and Research Synthesis: Theory, Value and Linkage" at the meeting of the American Educational Research Association, New Orleans (USA), April 2002.

Clarke, D., et al. (2002). Mathematics classroom practice: The learners' perspective. In C. Malcolm & C. Lubisi (Eds.), *Proceedings of the tenth annual meeting of the Southern African Association for*

Research in Mathematics, Science and Technology Education (pp. IV-7–VI-11). Durban: University of Natal.

Cole, M. & Means, B. (1981). *Comparative studies of how people think: An introduction.* Cambridge (USA): Harvard University Press.

Damerow, P., Dunkley, M. E., Nebres, B. F. & Werry, B. (Eds) (1984). *Mathematics for all* (Science and Technology Education Document Series No. 20). Paris: Unesco.

Damerow, P. & Westbury, I. (1984). Conclusions drawn from the experiences of the new mathematics movement. In P. Damerow, M. E. Dunkley, B. F. Nebres & B. Werry (Eds.), *Mathematics for all* (Science and Technology Education Document Series No. 20, pp. 22–25). Paris: Unesco.

Davis, E. (1987). Reasoning, commonsense. In S. C. Shapiro (Ed.), *The encyclopedia of artificial intelligence* (pp. 833–840). New York: Wiley.

Davis, E. (1990). *Representations of commonsense knowledge.* San Mateo (USA): Morgan Kaufmann.

Davis, P. J. (1996). Mathematics and common sense: Cooperation or conflict? In C. Keitel, U. Gellert, E. Jablonka & M. Müller (Eds.), *Mathematics (education) and common sense: The challenge of social change and technological development* (Proceedings of CIEAEM 47, pp. 29–39). Berlin: Free University of Berlin.

Davis, P. J. & Hersh, R. (1985). *The mathematical experience.* Boston: Birkhäuser.

Davis, P. J. & Hersh, R. (1986). *Descartes' dream.* San Diego: Harcourt.

Fischbein, E. (1987). *Intuition in science and mathematics: An educational approach.* Dordrecht: Reidel.

Forgas, J. P. (Ed.). (1981). *Social cognition: Perspectives on everyday understanding* (European Monographs in Social Psychology 26). London: Routledge.

Frankenstein, M. (1989). *Relearning mathematics: A different third R—radical maths.* London: Free Association Books.

Freudenthal, H. (1975). Pupils' achievements internationally compared—The IEA. *Educational Studies in Mathematics, 6,* 127–186.

Freudenthal, H. (1991). *Revisiting mathematics education: The China lectures.* Dordrecht: Kluwer.

Gattegno, C. (1974). *The common sense of teaching mathematics.* New York: Educational Solutions.

Geertz, C. (1983). *Local knowledge: Further essays in interpretive anthropology.* New York: Basic Books.

Gellert, U., Jablonka, E. & Keitel, C. (2001). Mathematical literacy and common sense in mathematics education. In B. Atweh, H. Forgasz & B. Nebres (Eds.), *Sociocultural research on mathematics education* (pp. 57–73). Mahwah (USA): Lawrence Erlbaum.

Glucksmann, A. (1987) *Descartes c'est la France* [Descartes is France]. Paris: Flammarion.

Greeno, J. G. (1991). Number sense as situated knowing in a conceptual domain. *Journal for Research in Mathematics Education, 22,* 170–218.

Habermas, J. (1971). *Toward a rational society.* London: Heinemann.

Harris, P. (1991). *Mathematics in a cultural context: Aboriginal perspectives on space, time and money.* Geelong (Australia): Deakin University.

Hoyles, C. & Sutherland, R. (1989). *Logo mathematics in the classroom.* London: Routledge.

Jablonka, E. & Gellert, U. (1996). Commonplaces about mathematics and the need for reflection. In C. Keitel, U. Gellert, E. Jablonka & M. Müller (Eds.), *Mathematics (education) and common sense: The challenge of social change and technological development* (Proceedings of CIEAEM 47, pp. 127–132). Berlin: Free University of Berlin.

Joseph, G. G. (1992). *The crest of the peacock: Non-European roots of mathematics.* London: Penguin Books.

Keitel, C. (2002). Students' struggle for sense making: Commonalities and differences in the search for meaning and significance in mathematics instruction. In K. D. Graf, F. Leung & F. Lopez-Real (Eds.), *Pre-conference proceedings of the ICMI Comparative Study Conference* (pp. 41–49). Hong Kong: University of Hong Kong, Faculty of Education.

Keitel, C., Kotzmann, E. & Skovsmose, O. (1993). Beyond the tunnel vision: Analysing the relationship between mathematics education, society and technology. In C. Keitel & K. Ruthven (Eds.), *Learning from computers: Mathematics education and technology* (pp. 242–279). Berlin: Springer.

Kline, M. (1985). *Mathematics and the search for knowledge.* Oxford: Oxford University Press.

Lave, J. (1988). *Cognition in practice: Mind, mathematics and culture in everyday life.* New York: Cambridge University Press.

Lenoir, T. (1979). Descartes and the geometrization of thought. *Historia Mathematica, 23,* 355–379.

Mason, J. (1996). Mathematical sense: Making the uncommon common and the insensible sensible. In C. Keitel, U. Gellert, E. Jablonka & M. Müller (Eds.), *Mathematics (education) and common sense: The challenge of social change and technological development* (Proceedings of CIEAEM 47, pp. 151–162). Berlin: Free University of Berlin.

Mason, J. & Monteiro, B. (1996). Conflicts between common sense and experience and concomitant shifts of attention. In C. Keitel, U. Gellert, E. Jablonka, and M. Müller (Eds.), *Mathematics (education) and common sense: The challenge of social change and technological development* (Proceedings of CIEAEM 47, pp. 97–102). Berlin: Free University of Berlin.

Maugham, W. S. (1949). *A writer's notebook.* Garden City, NY: Doubleday.

Mittelstrass, J. (1982). *Wissenschaft als Lebensform* [Knowledge as life form]. Frankfurt: Suhrkamp.

Moscovici, S. (1981). On social representations. In J. P. Forgas (Ed.), *Social cognition: Perspectives on everyday understanding* (European Monographs in Social Psychology 26, pp. 181–209). London: Routledge.

National Council of Teachers of Mathematics. (1989). *Curriculum and evaluation standards for school mathematics.* Reston (USA): Author.

National Council of Teachers of Mathematics. (1995). *Assessment standards for school mathematics.* Reston (USA): Author.

Newman, D., Griffin, P. & Cole, M. (1989). *The construction zone: Working for cognitive change in school.* Cambridge: Cambridge University Press.

Nissen, H., Damerow, P. & Englund, R. K. (1990). *Frühe Schrift und Techniken der Wirtschaftsverwaltung im Vorderen Orient: Informationsspeicherung und -verarbeitung vor 5000 Jahren* [Early writing and technologies of the management and administration of economics in the Near East: Information storing and processing 5000 years ago]. Bad Salzdetfurth (Germany): Franzbecker.

Paine, T. (1776). *Common sense; addressed to the inhabitants of America, on the following interesting subjects.* Philadelphia: R. Bell.

Pimm, D. (Ed.). (1988). *Mathematics, teachers and children.* London: Open University Press.

Pimm, D. (1991). [Review of the book *The mastery of reason*]. *Educational Studies in Mathematics, 22,* 391–405.

Resnick, L. (1987). *Education and learning to think.* Washington, DC: National Academy Press.

Restivo, S., van Bendegem, J. P. & Fischer, R. (Eds.) (1993). *Math worlds: Philosophical and social studies of mathematics and mathematics education.* Albany: SUNY Press.

Rogoff, B. & Lave, J. (Eds.). (1984). *Everyday cognition: Its development in social context.* Cambridge, MA: Harvard University Press.

Ruthven, K. (1996). Reconstructing common sense in mathematics education. In C. Keitel, U. Gellert, E. Jablonka & M. Müller (Eds.), *Mathematics (education) and common sense: The challenge of social change and technological development* (Proceedings of CIEAEM 47, pp. 107–112). Berlin: Free University of Berlin.

Saxe, G. B. & Gearhart, M. (Eds.). (1988). *Children's mathematics: New directions for child development.* San Francisco: Jossey-Bass.

Saxe, G. B. (1991). *Culture and cognitive development: Studies in mathematical understanding.* Hillsdale, NJ: Erlbaum.

Shimizu, Y. (2002, October). An analysis of lesson pattern in the sequence of mathematics lessons and associated values held by the teacher and the student. In K. D. Graf, F. Leung & F. Lopez-Real (Eds.), *Pre-conference proceedings of the ICMI Comparative Study Conference* (pp. 69–78). Hong Kong: University of Hong Kong, Faculty of Education.

Skovsmose, O. (1994). *Towards a philosophy of critical mathematics education.* Dordrecht: Kluwer.

Slavin, S. (1989). *All the math you'll ever need: A self-teaching guide.* New York: Wiley.

Solar, C. & Lafortune, L. (1994). *Des mathématiques autrement* [Mathematics in another way]. Montreal: Remue-Ménage.

Van den Heuvel-Panhuizen, M. (1994). New chances for paper-and-pencil tests. In L. Grugnetti (Ed.), *Assessment focussed on the student—L'évaluation centrée sur l'élève* (Proceedings of CIEAEM 45, pp. 213–222). Cagliari (Italy): University of Cagliari.

Vico, G. (1984). *The new science of Giambattista Vico* (T. G. Bergin & M. H. Fisch, Trans.). Ithaca, NY: Cornell University Press. (Original work published 1774)

Walkerdine, V. (1988). *The mastery of reason.* London: Routledge.

Weizenbaum, J. (1976). *Computer power and human reason: From judgment to calculation.* San Francisco: W. H. Freeman.

Wistedt, I. (1994). Everyday common sense and school mathematics. *European Journal of Psychology of Education, 9,* 139–147.

Zaslavsky, C. (1987). *Math comes all alive.* Portland (USA): J. Weston Walch.

COMMUNICATION AND CONSTRUCTION OF MEANING

Construction of mathematical meaning often results from processes of communication. In this introduction, we identify and discuss conditions and aspects of communication which may influence mathematical meaning. For example, meanings constructed between mathematicians are rather different from those of students on account of their different goals and concerns. Mathematicians are concerned with solving new problems while students are concerned with learning. Similarly, authors of textbooks are concerned with consistency of presentation and uniformity of approach while students look for convincing arguments, neat explanations and strategies which are economic in getting their task completed.

The nature of a community is not however the only determinant of mathematical meaning, which is a function of many factors emanating from diverse sources. The following vignette of a learning situation serves to illustrate this point.

> In the framework of a master class, in front of a public of mathematics educators, two 16-year-old students were asked to solve a geometrical task using dynamic geometry software. The situation was set up on a computer and can be described as follows: *ABC is a rectangular triangle with the right angle at A, P is any point on BC and D and E are the orthogonal projections of P onto AB and AC respectively.*
>
> The students explored the diagram by dragging the triangle and noticed how the relationships remained unchanged—that is that *A* was always a right angle, and *PD* and *PE* were always perpendicular to *AB* and *AC*. They were then asked to focus their attention on *P* and on *DE*. They were shown how to measure the length of *DE* and to display this value on the computer screen. Finally, they were asked to move *P* along *BC* and notice that the length of *DE* changed. A question was then posed to them: *What must be the position of P on BC in order for the length of DE to be minimal?*
>
> The students dragged *P* up and down *BC* watching how *DE* varied. They found a zone, a small segment of *BC*, minimising *DE* but were not able to characterise the location of *P* geometrically.

> The teacher suggested that they might be able to find a shape that was invariant under their transformations; they soon "saw" (without explicit justification) that *ADPE* was always a rectangle. They could then call upon the properties of a rectangle to help them—again after a nudge from the teacher—and suggested that the diagonals *DE* and *AP* must be the same length. It was a difficult step for the students to move from the statement of this property to its use in helping them to solve their problem but eventually they realised that they could replace the problem of minimising *DE* by that of minimising *AP*.
>
> But there was still work to do. How could they find the position of *P* which gave the smallest length for *AP*? By moving *P* about again and watching the measurements, they became convinced that *P* had to be at the foot of the perpendicular line from *A* to *BC*. But why? Their explanations were made exclusively in terms of the data from the measurements of *AP*: it was the correct position for *P* because when *P* was moving from this point in either direction, to the right or to the left, the length of *AP* increased. But the teacher was clearly not satisfied and asked for more reasons as to why they knew that *P* must be on this perpendicular line from *A*. One of the students whispered: "The shortest distance between two points is a straight line". At first sight this appeared to be a rather irrelevant statement but, perhaps surprisingly, it was accepted by the teacher. She was convinced that the boys had in fact understood the "correct" meaning and had simply failed to convey this to her in a precise language.

This vignette illustrates the complexity of communication patterns that may affect the construction of meaning. The aim of this introduction is to distinguish some of the conditions and aspects of communication that determine what mathematical meaning is constructed in teaching/learning situations. We will seek to elaborate the different influences and to examine how they interact. The introduction and the papers that follow achieve some coherence through their common focus on institutionalised teaching/learning situations and on the cognitive aspects and meaning. From the vantage point of these twin perspectives, a range of questions have been identified as being particularly germane to the factors of communication which affect meaning construction.

1. AGENTS AND THEIR STATUS: WHAT ARE THE AGENTS IN THE COMMUNICATION?

Communication during the teaching and learning of mathematics involves human and non-human agents: the teacher, students, text, software and other artifacts. These agents may be actually present (a student working with a text) or evoked by the human agents in their interaction as when students solving a problem in a text evoke some computer images or start discussing their solution in terms of the expectations of their teacher.

In the above vignette, the agents of communication are the pair of students, the teacher, and the geometry software. The teacher's aim is for the students to explore the geometric situation with the software and to solve the problem that was posed. The teacher's communicative actions are those of a facilitator; they are related to the practical aspects of using the software, to evoking the students' previous geometric knowledge, to giving hints and nudges. The students, on their part, have tacitly agreed to play along and they attempt to solve the problem. Their communicative acts seem to be task related.

In general, the human agents' communicative actions are likely to depend on their aims; a teacher may intend that students deal with a certain task because of some learning objectives he or she has in mind, and he or she may present this task to a class or individual student in a particular manner because of his or her assessment of the students' present needs. The students, on the other hand, are likely to react to the task in a way that depends on their own aims which may range from a deeply felt need to understand a specific reasoning pattern to the wish to get the task over with as quickly as possible.

Alongside the human participants, artifacts may play a pivotal role in structuring the interactions in teaching/learning situations—by the way they are designed and the tools made available. In the vignette for example, the dynamic portrayal of the problem made possible by the software may have crucially influenced the students' interactions in the task—by being able to drag *P*, they could more easily distinguish what changed and what remained invariant, and this may have led them to notice the invariant rectangle in the drawing. Texts also influence interactions in teaching/learning situations. In some scenarios they are indeed the dominant influence on the interpretation of the mathematical content of any curriculum and on the perceptions students might construct of mathematics and of themselves as mathematical learners.

Each of the interacting agents carries a particular status, explicit or implicit with respect to the knowledge being communicated. The status may be one of absolute authority, of a resource, of a moderator or a validator. In the vignette, the teacher acts as a facilitator rather than an authority. She poses questions rather than gives answers. The software, on the other hand, has a more authoritative status. The students don't question the accuracy of the pictorial and numerical outputs, until provoked by the teacher, and use these outputs as base for their solution to the problem. In fact, one of communicative acts of teacher was meant to put the authority of the software into question by showing that the measurements of the segment *DE* are not so accurate.

We see that agents may change in the course of the interaction as can their prominence and status. The students in the vignette initially relied completely on feedback from the computer. But with a nudge from the teacher they refocused on their mathematical knowledge to resolve the question of when the segment *AP* is the shortest. Such shifts in the role of agents affect both the way and the kind of meaning which is constructed.

2. THE MATHEMATICAL REFERENT: WHAT IS THE (INTENDED) MATHEMATICAL REFERENT OF THE COMMUNICATION?

In mathematical communication, the referents usually include one or several mathematical notions; they may also include notions from the applied problems which these mathematical referents allow to solve or from tools used to act on the mathematical referents (e.g., LOGO, Cabri-geometry). The task described in the vignette involved the use of several mathematical concepts: rectangle, diagonals, function, distance, minimum. But mentioning them only by such labels does not reflect how they had to be used, to what questions they gave an answer. For example, in the last step of the solution process, perpendicularity is used to minimise a distance, whereas in the first step it is used to characterise a rectangle. The former use links geometrical aspects of the problem to numerical ones whereas the latter is internal to geometry. Under the numerous properties of the rectangle, the congruence of the diagonals was the critical one for solving the problem. One of the key features of the solution lies in the interplay between geometrical properties such as perpendicularity, and magnitudes of geometrical objects. But again the numerical aspects of the solution referred only to comparison between lengths (bigger, smaller, equal) and not to the actual values of each of them. This certainly was, as mentioned later, a difficulty for the students who focused more on the values of lengths than on relations between them.

This shows that describing referents, in particular mathematical referents, by only one label (such as vector, function, reflection, etc.) fails to do justice to the mathematical content involved in the communication, content which must be described more precisely within mathematics itself.

Mathematical notions do not stand on their own but are determined by a set of relations with other notions. Their description must refer to the other mathematical concepts to which the content is related in the specific situation which is the object of study. The description must also take into account the type of use of the mathematical content which is intended. Or in other words, what kind of mathematical problems does it allow to solve? A notion like reflection may be involved in tasks of drawing the images of figures or finding the symmetry line of a given figure. In this latter case it is the object of the problem but it also may be a tool in problems of finding the locus of points or in some construction problems of objects satisfying several conditions. When describing the mathematical referent of the communication, one must take into account its interrelations with other notions as well as the uses and practices attached to it.

Moreover, the uses and practices and the problems faced by mathematicians are different at different junctures in history, thus leading to changes of the mathematical referents with time. This has been shown in well known historical analyses such as the one by Lakatos on Euler's formula for polyhedra or the one by Kleiner on functions.

3. MEANS AND MODES OF EXPRESSION: WHAT ARE THE LINGUISTIC AND THE SYMBOLIC MEANS USED IN COMMUNICATION?

Mathematical notions can be expressed in a variety of ways. A function, for example, can be given in a manner that is situation bound (the time dependence of the height of an object that was thrown upward from the roof of a 13 meter high house at a speed of 8 m/s), by means of a formula ($f(x) = 13 + 8x - 5x^2$, $0 < x < 2.6$), diagram (graph), picture (house, child, ball, indication of initial height and speed), verbally, or numerically. Each of these descriptions suggests different aspects of the mathematical referent in question; each of them is useful for different types of actions or operations on the function, and none of them is sufficient by itself to describe the function in such a way as to answer all questions that could be asked. But they all describe the same dependence between two variables and the inherent properties of this dependence (for example, the fact that the values of $f(x)$ increase for $0 < x < 0.8$) do not change when passing from one description to another. Thus the essence of the function resides not mainly in the descriptions used to represent it but rather in the properties of the mathematical referent, in the links that can be established between the descriptions, in what remains invariant under transition from one description to another.

More generally, any attempt to refer to a mathematical notion necessarily makes use of some means of expression, often language; any communication about mathematics, in particular communication during teaching/learning processes, uses specific formulations and settings which may have a strong influence on the mathematical content. For example, the "same" problem presented to the students in the vignette would presumably have taken on quite a different meaning if it had appeared in a different formulation ("Find the minimal length for *DE*, where...") and been posed to students with some background in calculus and analytic geometry. In summary, the meaning of a mathematical notion or problem depends on the symbolic and linguistic means used in communication about it and on the situation in which it appears.

In addition, gestures and tone of voice of the human agents involved in communication may also contribute to the meaning that is associated with mathematical notions or actions, just like in any other human communication. We don't know why the teacher in the vignette was convinced by the students' inadequate justification that the shortest distance between two points is a straight line; but we can assume that tone of voice or gestures may have been at least partly responsible.

4. INSTITUTIONAL MODALITIES: WHAT ARE THE INSTITUTIONAL AND MATERIAL CIRCUMSTANCES OF COMMUNICATION?

The lesson described in the vignette took place in slightly unusual circumstances. It was during a demonstration of a master class in a large room and in front of an audience. But, not unlike a classroom situation, time constraints were a factor in the communication. The teacher was obviously aware of the time factor, and her hints to the students may have been more direct as a way to move the session along. The fact that there was an audience may have also affected the students' way of communicating with each other, making them more guarded about what they were saying.

More generally, when considering the factors which affect how meaning is constructed within a sphere of institutional practice, we cannot ignore the modalities of the institution and how they affect and constrain communication. There is a curriculum to be covered, classes and lectures are schedule-bound, certain times and forms of assessment may be prescribed, classes may be crowded, and there are material considerations such as the availability of computers and other resources. Schools may also limit the kinds of mathematical situations which are communicated to those which they feel are consistent with the school culture.

5. WAYS OF KNOWING: WHAT ARE THE PROTAGONISTS' WAYS OF KNOWING?

When we analyse the role of the agents in any teaching/learning situation, we not only have to describe them at the time of their interaction but also have to take into account their individual histories. Can we tease out any characteristics of the developmental history of any agent that might be interpreted as impinging on the nature and pattern of the interrelationships between the agents at the particular time of the learning situation under study? Are there features of an agent's "way of knowing" that influence the interactions and the way these interactions are interpreted?

Let us first consider the part played by the teacher. We know that teachers' knowledge of and beliefs about mathematics will affect their interactions with students. In the vignette, the teacher clearly sees that it is important for students to explore the geometrical situation and to attempt to construct meaning for themselves; yet she has a clear agenda with a well-defined end point to the investigation. Similarly, the students come to the learning situation with a host of prior experiences in mathematics from which they have developed knowledge but also expectations of the teaching situation. The students in the vignette for example whilst happy to explore the problem of minimising DE were clearly uncomfortable with the geometry of the situation. They did not readily exploit the properties of the rectangle as tools to assist them in solving the problem.

For teachers and students, there are further sets of factors that we wish to consider as influencing their mutual interactions and their interactions with the artifacts and texts and thus their ways of knowing. The first set concerns their judgments as to how far the outcome of any interaction has met the objectives of the activity. Clearly certain criteria of validity flow from the mathematical nature of the activity but these are always mediated by the agents in the interaction. Ultimately it is the protagonists who decide: for the teacher this decision will play a role in structuring and planning instructional strategies; for the students it will influence their expectations and engagement in problem situations in the future.

A second set of factors relate to the habits built up over time that help to describe and to explain consistencies identifiable in teaching/learning situations in mathematics. These consistencies have a behavioural component: the working habits that comprise what is stressed and what is ignored in the teaching/learning situation; the practices followed almost as a matter of routine. But there are also habits that reside deeper: the habits of mind of the protagonists, their expectations of any interaction with other agents and indeed with mathematics as experienced in any particular sphere of practice.

As illustration in our vignette, the teacher, in contrast to the students, is unhappy with a solution based solely on measurement and indeed refused to accept these data as valid justification. At first, the inclination of the students was to close the problem at the point when they were convinced that they had identified the position of *P*. They expected that they had solved the problem and were only moved to articulate a geometrical justification—albeit tentatively—in the face of the teacher's persistent questioning. The students' reluctance was evident in the hesitant, even confused way they eventually came up with a geometrical solution. Again, the teacher's satisfaction at this was apparent even to the point that she was willing to accept as valid their rather imprecise final formulation.

6. WAYS OF INTERACTING: WHAT ARE THE ACTORS' (EXPLICIT OR IMPLICIT) WAYS OF INTERACTION?

In goal-oriented, content-focused communication situations such as those arising in education, the agents of communication quickly develop certain ways of interaction that make these situations a little more predictable for them. This means that the interactions are subject to rules or fall into routines even if these rules and routines are not explicitly stated or consciously chosen.

Sets of rules and routines of (purposeful) interaction between people in definite social roles (e.g., an infant and a mother, a student and a teacher) constitute what Bruner called "formats of interaction". One could stretch this notion to encompass interactions not only between people but also between a person, and, say, a text or a software program. An important issue in mathematics education is how particular patterns or "formats of interaction" affect what students understand and learn in and of mathematics: what kinds of mathematical meanings do they construct in one format as opposed to another format of interaction. For example, how does the way in which a student decides to work with a text influence what sense he or she makes of the mathematics this text intended to teach him or her? What are the means of control that a student develops over the feedback from a computer software? How do the protagonists in a mathematical communication make sure they understand each other? For example, in checking a student's understanding of a task, does the teacher interrogate the student about the definitions of the key words, or does the teacher allow for tentative interpretations to be produced and react by challenging questions or counterexamples? It seems that in each case, the student will put to work a different way of knowing and therefore, different meanings will be constructed as a result of the communication. On the other side, the student can use different strategies, direct or indirect, bold or shy, confident or lacking in confidence, to ascertain that he or she got the teacher's idea. The student may ask: "Is that what you want me to do?", or say: "This does not make sense to me. What you are saying implies that... etc. etc. Is that what you mean?" Again, the two types of questions trigger different types of routines of interaction which lead to the construction of very different types of mathematical meanings.

The vignette at the beginning of this introduction tells us very little about interactions between the teacher and the students. What we learn, however, raises several questions about these interactions. Each party—the teacher on the one hand, and the students on the other—seemed to be trying to establish a routine of interaction based on their own interpretations of the other party's intentions. These interpretations may have been erroneous and the course of communication could perhaps be explained by a certain tension between two routines of interaction.

We are told in the vignette that the students were shown how to move the point P on BC and how to have the measures of DE displayed on the screen. We are then told that, in explaining why AP must be perpendicular to BC, they refer to the data from the measurement of AP as P moved along BC. We also learn that the teacher was not satisfied with this explanation and showed signs of approval only when she heard one of the students stating something that sounded like a geometric definition or theorem.

The routine of interaction that the teacher was trying to establish could have been the following: She first gives the students the necessary tools for guessing the answer to the problem. The students use the tools and indeed make the expected guess. By finding the answer for themselves, they become "the owners" of the problem. The teacher expects now that, when prompted to geometrically validate their guess, they will endorse this task as their own responsibility. She believes that, at this moment, the "handover" (devolution) of the task has occurred, that the students are not trying to satisfy her presumed expectations but attempt to prove to themselves that they are right. To her, measurements alone do not provide sufficient evidence for the conjecture to be true. After all, with the 2-digit approximations, there is a whole segment of points that corresponds to the same "minimal" measurement of AP. She expects the students to think similarly: If they wanted to behave as empiricists they would fix one of the points for which the measure was the smallest, and ask the software if, in this case, AP is perpendicular to BC. It is very likely that the answer would be: "Objects are not perpendicular". For the teacher, it is quite obvious that the students would not regard this empirical evidence as a refutation of their conjecture. They will thus abandon their empiricist position and turn to theory in search for proof.

Let us speculate now how the situation could look from the perspective of the students: When they are shown how to move points on the screen and have measurements of line segments displayed, the students believe that this is "the lesson" or "the new stuff" that they are being taught. Moreover, they know from their experience at school that, when something new is being taught, usually the exercises that follow are meant for the students to practice this new knowledge. Thus they think that "using measurements" is what they are expected to practice, and this is what they do, "moving *P* about and watching the measurements", and giving explanations in terms of measurements. To their surprise, however, the teacher rejects their explanations; she is not satisfied with them. They feel some sort of breach at this point: Either the teacher has broken the implicit "contract" by abruptly changing the routine she seemed to be previously following, or they have totally misunderstood the rules of the game. In any case, they feel the problem is out of their hands, and just try to figure out what is expected of them. If measurement, the "new stuff" is no good, maybe one has to refer to some older knowledge. One of them recalls a statement that had something to do with minimality: "The shortest distance between two points is a straight line", and produces it. The statement does not explain anything with regard to their conjecture but, surprisingly, the teacher accepts it, and closes the episode as if the problem was solved. For the students the problem is not solved; it only begins just now.

We do not know if things indeed happened this way and, in fact, this question is beside the point. What we want to illustrate by the speculations above is how difficult is the process of establishment of routines of interaction that would lead to some form of *shared understanding* and render the situations of communication a little bit more *predictable*. We also want to suggest how such studies of teacher-student interaction naturally lead to the question: What are the understandings that the students develop in the course of their interactions with teachers? For example, in our case: What sense have the students made of the use of the software in doing geometry after the episode referred to in the vignette? They were sitting in front of the computer, they were using it to find a solution, but they were not allowed to reason in terms of the software. What did this mean to them?

7. CONCLUDING REMARKS

In this interlude various aspects of communication which may affect the construction of meaning are discussed. On the other hand, the problem of construction of meaning itself is not really tackled. This is an evasive problem: It is difficult to know what each partner thinks; we can only hypothesise this by interpreting what they do and say; one of these hypotheses is that it is very likely that the interpretations of students are different from those of the teacher. These various interpretations of the situation are what we call meaning in the introduction.

CELIA HOYLES

MAKING MATHEMATICS AND SHARING MATHEMATICS: TWO PATHS TO CO-CONSTRUCTING MEANING?[1]

This chapter aims to explore the idea of computer use as a window on meaning construction. It takes constructionism and socio-cultural theory as starting points, and while recognising some of their internal contradictions, nonetheless seeks in a dialectical way to draw on their complementary ideas in order to develop a framework for discussing how meaning in mathematics can be constructed and shared through interaction with computer tools. Thus I am interested in analysing how students use computer tools to solve problems, and how during this process, their conceptions of the mathematical ideas involved are both externalised and mediated. Constructionism shares with constructivism the notion that student conceptions are resources on which mathematised notions can be built (see Smith, diSessa & Roschelle, 1993). But constructionism adds to constructivism the idea that: "this happens especially felicitously in a context where the learner is consciously engaged in constructing a public entity" (Harel & Papert, 1991). At the heart of constructionism is therefore the idea of "learning by making", where the object that is made is available for inspection and modification by the learner. Constructionism thus draws attention to the tools used in construction, the technology, not only how the tools constrain what can be built, but also, as a mathematics education researcher, how far the tools are appropriate for the task in hand from a mathematical point of view. Socio-cultural theories of mediated action complements a constructionist approach by assisting in interpretation of how the tools might be conceived by the students while they collectively negotiate their plans and their implementation strategies (see Wersch, 1991).

In this chapter, I will argue that the mediation of student conceptions by interaction with software tools that form part of a carefully designed exploratory environment has the potential to orient students towards a mathematical way of thinking, although not necessarily precisely in the direction planned by the teacher. Additionally, I will seek to show how the computer tool kit can serve as a shared concrete resource, a joint problem space, to manipulate and to mathematise.

1 I wish to acknowledge the contribution of my colleague Lulu Healy, with whom I undertook the case studies reported in this paper.

1. DESIGNING MATHEMATICAL MICROWORLDS

My concern is the iterative design and evaluation of mathematical microworlds. Perhaps the term microworld needs a little qualification at this point: the word has come to connote almost any exploratory learning environment that incorporates a computer. But the etymology of the word reveals an important meaning which has, perhaps, become a little clouded (for further elaboration, see Hoyles, 1993): a microworld was conceived as a world which was simultaneously rich and simple enough to study learning behaviour (originally, of machines); and it is this meaning as well as the more conventional one which I reassert here. An overriding principle for microworld design can be extrapolated: the need to build an environment where, potentially, there is a dialectical relationship between action and mathematical meaning through the mediation of the software tools. Put another way, a microworld consists of tools which are designed to connect to the students' points of view and, by providing a medium through which mathematics can be communicated, can also orientate them towards a mathematical perspective; or in the language of the introduction to this section, the microworld is an interacting agent which specifies the means by which mathematics will be expressed.

Herein lies a central assumption, which is that meaningful and general mathematical relationships can be constructed and articulated even by those with little prior access to the semantics and syntax of conventional representational infrastructures for mathematics (see Kaput, Noss & Hoyles, 2002 for a discussion of this idea). How might this work? I conjecture that an answer might relate to the provision of carefully designed and integrated tools, activities and teacher advice (see Hoyles, 2002 for a discussion of the need to pay more attention to design in mathematics education research within the socio-cultural paradigm). Together the learning system (a term coined by Healy, 2001) comprising these three components should offer, on the one hand, a structure of local support that is contingent on the state of the learner's current mathematical understandings at any one time, and on the other, the possibility of building an emergent global support structure from the connections forged in use by the learners who collectively form the "mathematical" community of practice (Lave & Wenger, 1991).

In order to characterise this dual support system, I will adopt the metaphor borrowed by Noss & Hoyles (1996) from a current technology, the World Wide Web, and call this expanding, interconnected, network of ideas, each one built by an individual (or group of individuals), yet drawing on the resources available at the time of building—a web. Our notion of webbing aimed to synthesise Vygotskian and Piagetian approaches to learning, by arguing that there can be connections built into the structure of any environment, signposts which assist in navigation, yet the signposts followed and the connections reconstructed for support stand in a dialectical relationship with the actors in the system. Following Wilensky and others (see for example, Wilensky, 1993), we also proposed the notion of situated abstraction as a process of connection to new objects, a process that develops in activity: abstracting within a domain rather than away from it. Situated abstraction is both process and object: an "articulation", a "statement", but also a (re)thinking-in-progress, situated and shaped by the expressive tools at hand and the community who use them. A situated abstraction is thus a "persons/tools" conceptual construct underscoring the idea, first that the condition of a mathematical concept "being abstract" does not come ready-made, either a priori or post hoc, and second that the genesis of a situated abstraction is constitutive of its meaning, that is traces of the tools, social norms and activities that scaffolded the learning remain as an integral part of the mathematical concept (for further elaboration of these ideas, see Noss & Hoyles, 1996; Hoyles, Noss & Pozzi, 2001; Noss, Hoyles & Pozzi, 2002).

Although webbing and abstracting are complementary, this complementarity is sometimes hidden. In many settings, mathematical invariants underpin actions and learners exploit these in functional ways to achieve their goals. Yet these invariants remain unexpressed, implicit in the action, and may be recognised only by an observer versed in mathematical ways of thinking. By contrast, in microworlds, some of these invariants though rooted in action are also articulated—quasi-mathematically—in the operational terms of the available tools. Learners construct situated abstractions of mathematical ideas through actions on these tools, through reflecting on what they have done and communicating it to others—a process which extends the ideas but also shapes them.

Vygotsky's ideas are useful in that they move the focus of attention from mathematical objects to the dialectical relationship of action on the objects and thought (see for example, Vygotsky, 1986). Additionally in a Vygotskian perspective tools are mediators, intermediary socio-cultural agents intervening in social transactions—a crucial part of the dual process of internalising-externalising (see Vygotsky, 1978). Wersch (1991) similarly emphasises the role of mediated action: "human action typically employs "mediational means" such as tools and language and these mediational means shape the action in essential ways" (p. 12). From this point of view, learning mathematics is not simply a matter of abstracting a certain structure or form from an activity, but a more complex process involving actions, representations and social practices. Simply looking at invariants identified through reflection on actions is only part of the story: mathematical meanings are inextricably interwoven with the tools used in their construction and with the way these tools represent mathematical invariants and express relationships. Socio-cultural theory can provide insight into this mutual shaping of student/computer interchanges. It also brings to the fore crucially important strategic questions for design. First the design of the tools: what should be their "level of abstraction"? Or, put another way, how much should be "done" by the software and what should be left for the students to construct? Second the design of the activities in which the tools are embedded: what tasks should be developed and classroom social norms encouraged that foster students' engagement with mathematical ideas and discourse? In this chapter, my focus is on the former question and the role of the tools. The tools must do "just enough" to illuminate structures and relationships while not solving the task completely, and I shall seek to illustrate the importance of finding an appropriate balance between these two poles if both student and mathematical meanings are to be simultaneously respected.

There are many influences on the way software tools are used, such as the demands of the task and the experiences and goals of the learner and the classroom mathematical community. Just as software tools mediate students' ideas through action and representation, the social setting also shapes the learning process by the range of viewpoints brought to bear on the activities and the ways these are communicated. We have to look beyond the notion of an individual constructing his or her own knowledge towards a consideration of the social framework within which activities take place and how social interaction transcends and transforms individual conceptual structures. Microworld design therefore includes planning for active encounters between students so they can construct a joint cognitive system; so students together with the software co-construct mathematical knowledge through experimentation and social engagement.[2]

Placing co-construction or emergent consensus as the motor for learning appears at first sight to stand in contradiction to a Piagetian viewpoint that stresses the centrality of conflict and socio-cognitive conflict: the ability to stand back, decentrate, and reflect upon one's own activity in the light of differing points of view. Yet this is not necessarily the case. Many peer collaboration studies suggest that for learning to occur there is a need to exchange diverse views over an extended period of time and to strive to achieve a higher synthesis (see for example, Kruger, 1993). In line with this framework, our design of microworld activities attempts to configure conflict and co-construction as two sides of the same coin rather than in opposition; it seeks to provoke conflict (from computer feedback, for example), but also to encourage the negotiation of joint action and resolution (see Hoyles, Healy & Pozzi, 1990; Healy, Pozzi & Hoyles, 1995 for description and analysis of this approach in the context of groupwork with computers).

Incorporating co-construction into the microworld agenda has implications for the design of microworld tools. Clearly the software will constrain action but it must do more than this: it must provide ways to "disperse" some of the load of the task (see Crook, 1994), and it must serve as a mediator of social interaction, a medium through which shared expression can be constructed (for further elaboration, see Confrey, Smith, Piliero & Rizzuti, 1991; and Roschelle, 1992). In fact the idea of webbing emerged during our struggle to reconcile the individual's role in the construction of mathematical meaning with the recognition of the part played by social and cultural forces. But mathematics educators have a more complex problem to face than social psychologists. Their goal is more exacting than the achievement of "any" negotiated consensus, since there is a need to be able to recognise in any consensus that the activity is mathematical and so must conform to prevailing mathematical norms.

I shall seek to show in the sections that follow how it is possible for students to use and interrelate computer tools in ways that forge links between their worlds and the mathematical world of their community, and in the process of developing their

[2] It is interesting to trace the origins of the now rather popular notion of co-construction. The first sighting of the word I can find is in Forman, who largely sets it up as a contrast to conflict: On page 37 she describes a situation: "co-construction versus social conflict" (Forman & Kraker, 1985).

situated meanings change both. I will explore the process of abstracting in situ through some case studies of students investigating geometry problems. The choice of mathematical referent as geometry has of course implications for design, so I introduce the case studies with a brief summary of the way geometry has been approached in England and Wales.

2. SCHOOL GEOMETRY IN ENGLAND AND WALES

Students come to the study of geometry with strong intuitions of shape and space drawn from their day-to-day experiences. As Freudenthal (1973) has argued: "Geometry is one of the best opportunities which exists to learn how to mathematize reality" (p. 407). Put another way and drawing on the terminology of some of the chapters in this book (e.g., Keitel & Kilpatrick), geometry offers a unique opportunity for building connections between commonsense notions picked up as part of everyday activity and their formalisation within mathematical discourse, and, given the dual nature of geometry, injecting axiomatic systems with visual meanings.

This balance between visual intuition and formal deductive reasoning is however rarely achieved in a school curriculum. In the past, in England and Wales at least, the geometry curriculum was dominated by axioms, definitions, theorems and corollaries, exercises in "pure" deductive reasoning with little attempt to "connect" the mathematical objects at the heart of the proofs with students' spontaneous intuitions about their visual world. More recently, in contrast, the deductive and axiomatic side of geometry all but disappeared. Students named shapes and were encouraged to explore some of their properties, but engaged in few ruler and compass constructions and little or no deductive proof.[3]

Some appreciation of the curriculum followed by the 12 year old students in the case studies that follow is important in trying to understand and interpret their responses; for example, much of the basic vocabulary of geometry (perpendicular, intersection, bisector) would have been unfamiliar. However, the students would have been inducted into an investigative culture in mathematics and were accustomed to conjecturing and exploring for themselves. The particular mathematical topics that are explored or used as part of the microworld investigations reported are reflection and symmetry. Students would have come across these notions in primary school through a range of practical activities and they are revisited in later years. The programmes of study in Key Stage 2 (ages about 7–11 years) and Key Stages 3 and 4 (ages 11–16 years) of the National Curriculum for England and Wales in operation at that time and specifically relate to these topics are shown in Table 1.

[3] The curriculum was revised in 1999, when geometrical reasoning with some elements of deduction was introduced.

Key Stage 2	Transform 2-D shapes of translation, reflection and rotation, and visualise movements and simple transformations to create and describe patterns;
Key Stages 3 and 4	Recognise and visualise the transformations of translation, reflection, rotation and enlargement, and their combination in two dimensions; understand the notations used to describe them;
	Understand and use the properties of transformations to create and analyse patterns, to investigate the properties of shapes, and to derive results, including congruence.

Table 1: Programmes of Study covering reflection and symmetry in the Mathematics National Curriculum of England and Wales (DFE, 1995)

The National Curriculum was also organised into levels of attainment and the performance indicators that students at a particular level are expected characteristically to demonstrate are also specified[4]. Relevant sections are shown in Table 2.

Level 3	Pupils classify 3-D and 2-D shapes in various ways using mathematical properties such as reflective symmetry.
Level 4	They reflect simple shapes in a mirror line.
Level 5	They identify all the symmetries of 2-D shapes.
Level 6	They solve problems using angle and symmetry properties of polygons and properties of intersecting and parallel lines, and explain these properties.

Table 2: Performance indicators related to reflection and symmetry

Thus in the curriculum, reflection was closely connected to ideas of symmetry. This notion (with which children are familiar from everyday usage) is introduced through practical activities, such as folding paper or using mirrors, and later extended (at around age 12 years) by requiring reflections of objects drawn on grids; a typical example is shown in Figure 1. Thus students are expected to know what a reflection looks like and develop a qualitative appreciation of the transformation after which they are considered to be in a position to have an appropriate understanding of the relationships involved. There is little, if any, explicit discussion of the properties of "mirror" images.

[4] At Key Stage 2, the majority of students should be in the range 2 to 5, and by the end of Key Stage 3, within the range 3 to 7 (the scale does not apply to Key Stage 4).

Complete this design by drawing the reflected shape.
You cannot *fold* or use a *mirror*.

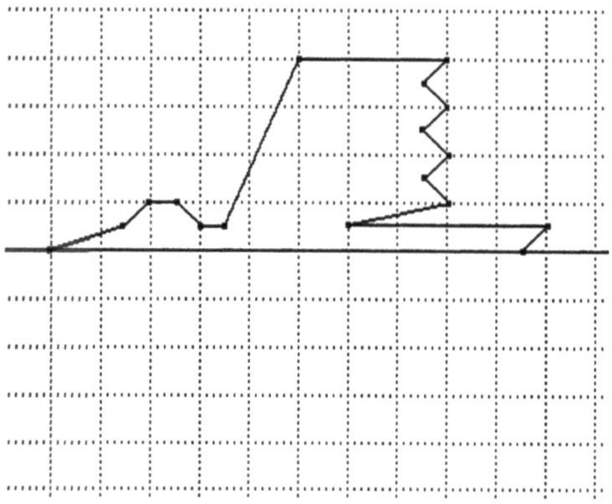

Figure 1: A typical school activity about reflection

Another feature of the curriculum with regard to reflection is that it appears in isolation from other mathematical structures: It is not studied as an example of the group of isometries, nor is it considered as a transformation of the plane, a function to be manipulated and combined with other transformations. It is simply an activity that is an end in itself. Thus the implementation of the curriculum seems to convey two meanings for reflection: it is an action performed on an object to produce an image, although the exact nature of the action is not made explicit, and it is connected to symmetry.

So how do students view reflections? In a survey of students' ideas of reflection which took place in the same school as the case studies to be reported, we interviewed 50 or so students of the same age (12–13 years) and found that they adopted a range of strategies that varied according to task features, but which all exhibited little if any appreciation of the angle relationships that had to be satisfied between an object, its image and the mirror if the transformation was a reflection (see Hoyles & Healy, 1996). Although most students were able to reflect correctly horizontal or vertical objects in horizontal or vertical mirror lines, they displayed a range of errors when these initial conditions were changed: for example, objects with a horizontal/vertical orientation were treated differently from those with a slanted orientation, complex objects differently from single points, and an object that crossed the mirror line differently from one that did not (see also Küchemann, 1981; Grenier, 1987). From an analysis of these responses, we concluded that the students appeared to operate in three distinct worlds depending on the presence of horizontal/vertical mirrors, horizontal/vertical shapes, or slanted mirrors and shapes.

Perhaps unsurprisingly given their previous teaching, the majority of students described reflective symmetry in ways simply derived from experiential meanings that were unconnected to any precise mathematical definition, and largely used pragmatic approaches (measuring or visual criteria) to check the validity of their constructions. Students probably appreciated at some level that reflective images must be the same distance from a mirror but how this was operationalised varied according to the orientation of the mirror and the relationship of object to mirror. Students also knew that these images comprised something "opposite", though the meaning of opposite was rather ill defined. These analyses were derived from responses to paper and pencil tasks. I now turn to two case studies of 13 year old students working in a computer microworld to try to tease out how the students' meanings of these geometrical notions were mediated and developed by interaction with the software tools.

It should be noted that in the case-study school, computers were part of the culture of the mathematics classroom: every mathematics classroom was equipped with at least one, usually, two computers. The children followed an individualised scheme and worked at tables in pairs and groups and one or more pair would always be working with the computer, using most frequently spreadsheets or Logo. Logo was considered to be an everyday part of a student's tool kit in much the same way as a calculator or a pencil and as such was viewed as a problem-solving tool in mathematics that students could exploit when they felt it was appropriate. Cabri had also been introduced in a fairly intensive way with support from a research team from the University.[5]

3. THE CASE STUDIES

Different software programs were used in the two case studies described here. The first comprised a specially designed microworld, Turtle Mirrors (TM), written in Microworlds Project Builder (MPB), a version of Logo; the second was Cabri Geometry, a dynamic tool kit for Euclidean geometry.

3.1 Designing a tool for constructing a reflection[6]

Prior to the episode described here, the students had been introduced to MPB.[7] We had considerable experience with previous versions of Logo and chose MPB as we were struck by the potential of its multiple turtles, object-oriented structures, and elements of direct manipulation (turtles for example could be picked up and moved, they could be "spoken to" and could communicate with each other). MPB also has simple interface features, buttons and sliders, which were rather transparent ways to help students control their own projects or generalise their strategies: pressing a button could run a command or a procedure and moving a slider could change a variable. For facilitating communication between students, MPB also has text boxes in which (among other things) messages can be written to explain ideas and record

[5] Comprising Lulu Healy, Richard Noss, and the author.

[6] All this work was undertaken collaboratively with Lulu Healy and Richard Noss.

[7] The transition from standard Logo to MPB is relatively unproblematic.

thoughts, comments or the findings of an investigation, and also in which a reusable trace of the history of a turtle's movements can be activated.

In MPB, there are no primitives that "do" reflections so we were faced with our first design decision. We could simply write tools that would reflect objects in mirrors at various orientations, programs whose "inner workings" would be "black" boxes for students, but into which could be input different objects and mirror orientations to obtain a reflective image. The focus of attention of the exploration would in these circumstances be on the visual relationship of image to object (an example of this approach is described in Edwards, 1992). We decided not to go down this road as we wished but to tease out how students might set about the construction of reflective images in a computer world of communicating turtles where there were no menu items or predefined tools for what they had to do. We then would provide them with tools so they could implement *their* preferred strategies, and this emergent microworld, which comprised the primitives of MPB and the emergent set of tools, we called Turtle Mirrors, or TM.[8] We anticipated that adopting this approach would help the students to focus on the process as well as the outcome of their constructions, and thus begin to appreciate the invariants and properties they had expressed using the tools.

We now turn to the work of a pair of girls to illustrate our approach. The girls had used TM to find a reflection of an object in a vertical mirror and had come up (jointly) with a construction that had evolved from an initially rather syntactic notion of reversing almost any commands through gradual refinements to end in a precise method involving switching right and left turns when constructing the image of an object under reflection. The pair had abstracted this general relationship between the process of construction of object and image, which was formulated in terms of the tools of the microworld ("when you see a RIGHT make it LEFT") worked for all reflections. But what would happen when the Logo code with its RIGHTS and LEFTS underlying the figures was taken away? Would the students continue to be able to exploit their situated meanings of reflection?

The pair opened a screen that showed an object drawn in blue by a blue turtle and its image under reflection in red, drawn by a red turtle, with these two turtles starting from the same initial positions relative to the object or its image (see Figure 2).[9] The task was to draw the missing mirror. On the screen were also text boxes (the rectangles with frames in Figure 2) in which symbolic histories of a specific turtle could be displayed (the contents of text box, `Blues`, for example, could contain a record of all the movements of the blue turtle). We predicted that reflecting on and reusing the contents of these text boxes would help students notice the relationship between the process of construction of object and of image simultaneously with comparing their visual expressions. Finally, we designed buttons on the screen that when pressed would run a command or procedure and thus facilitate the easy management of the work space: buttons were made to reset the page at any point

8 For details of the design of Turtle Mirrors, see Hoyles & Healy (1997). Healy (2002) has gone on iteratively to design a learning system for reflection using a multiple-turtle geometry microworld.

9 Since the figures are not in colour, the turtles have simply been labelled with their colours.

with a mirror at a different orientation, to clear boxes of redundant methods and to highlight and run commands in the screen text boxes. For example, in Figure 2, pressing the button, `do_blues` would activate the blue turtle and make it perform all commands in the text box `Blues`.

So what did the two girls do to find the mirror? Cheryl immediately picked up the blue turtle, placing it where she thought the mirror should be located and tracing with her finger a visual estimation of the orientation of the mirror line. Emily immediately accepted that visually this line seemed correct, but was unhappy with the imprecision of simply dragging a turtle to the required position and sending it off to draw the line in such an imprecise way:

Emily: No, it has to be, we have to do it in exactly the right place. [...] We have to work out how to get them to the right place.
Cheryl: Which is about there, I know, we haven't done it yet.

The girls knew that they had to construct the mirror line so that it conformed to their visual estimation and began to negotiate a strategy. Emily began to talk about angles, aware that they were important for reflections, but at the same time, was not entirely sure how:

Emily: Well if we worked out the angle, could we have... er, they are both going at an angle and this is at this angle. Could that help? [tracing the path of the red image on the screen]

As Emily started to follow the steps of the red turtle on the screen, Cheryl had an idea. She traced the path of *both* turtles simultaneously on the screen, starting from the top point, going round each A and continuing with both hands until two turtles would meet. Emily immediately took up this idea:

Emily: Do you want to do that? Then the mirror line is always going to be there, isn't it?

The girls appeared absolutely convinced that this point where the two turtles would meet must lie on the mirror, we conjecture because it satisfied their intuitions of symmetry. In our trialling of TM, we had noted that students often had wanted to "get equivalent turtles to meet" and so had written a tool, `homein`, which allowed this to be operationalised: typing `"red, homein "blue` or `"blue, homein "red`, sends the two turtles towards each other until they met. In this tool the turtles moved at the same speed, a relationship that was implicit in Cheryl's gestures but was not articulated. We showed `homein` to the two girls and they used it to locate a point on the mirror line as shown in Figure 3.

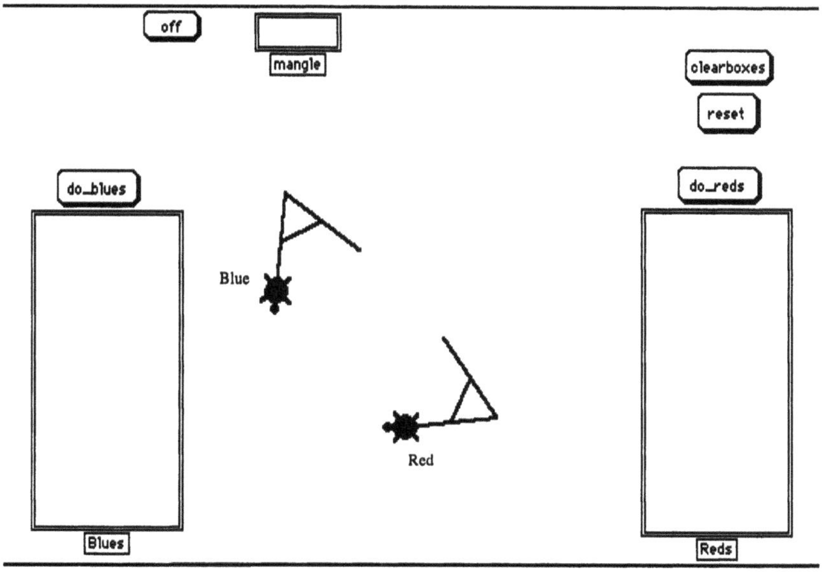

Figure 2: Find the missing mirror line

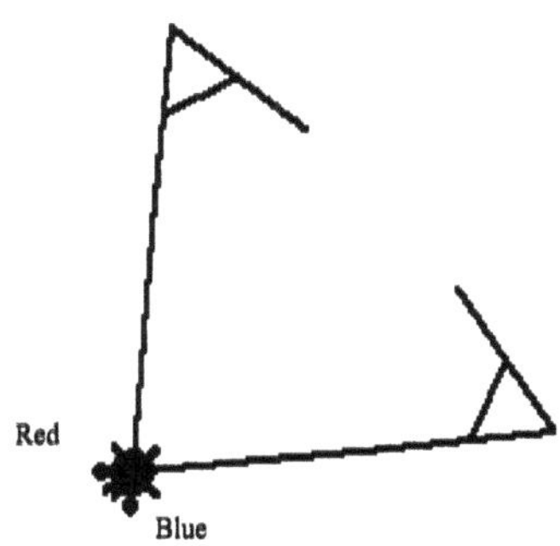

Figure 3: Using homein to locate a point on the mirror line

Emily and Cheryl now realised that they had next to find a way to turn the turtles so they faced in the direction of the mirror line. They could have used a `lineup` tool that we had written specifically for this purpose (it turned two turtles at the same speed until they had the same heading). But instead the girls devised their own method that called upon their visual estimation of the mirror's orientation, their insistence on the symmetry of the construction of object and of image and their left/right swapping strategy.

Emily:	Do we just want to turn it round?
Cheryl:	Mm, which shall we do red or blue?
Emily:	I don't think it... won't they be the same?
Cheryl:	Yes... oh, but left to right and that.

Given the pair did not know how far to turn the turtles, they decided on an iterative strategy. They decided to try to find an angle through they could turn the blue turtle in one direction and the red turtle in the other direction, so that both turtles would end up facing in the same direction. To implement this idea, they moved the blue turtle through a series of LT 10's, testing at each iteration its direction against their visual estimate of the mirror line. When they were satisfied it looked about right, they turned the red turtle through the same angle but to the right. When they discovered that it ended facing exactly the same way as the blue, they had confirming evidence that they had found the direction of the mirror line and were ready to draw it.

The vignette illustrates how Emily and Cheryl exploited TM's tools to satisfy their intuitions of reflection as involving symmetry. The intuitions became expressed in their work as the mirror line should be "in the middle", and there should be "equivalent" but right/ left switched turtle paths. The important steps in the pair's successful solving of the problem seemed to be: that the girls were able to share their visualisations of the required construction and the means to formalise them in terms of turtle traces; they could utilise their left/right reversal strategy and connect it to existing knowledge of mirror lines; and they could bring their visual and construction methods into a harmony where each served as a checking mechanism for the other.

Turning to the role of the tools, it is important to notice that wrapped into the tool `homein` was a feature designed to assist in thinking about symmetry, namely that the turtles moved at the same speed. It is not clear that the students were at all aware of this constraint, let alone its importance. How far students should notice a tool's mediating role is an important issue that has implications beyond this case study.

Returning to the case study, it must be mentioned that not all our students chose to use the same set of TM tools. Cheryl and Emily for example similarly started with a clear idea as to where the mirror should be but wanted to find "the middle" by using what they described as "equivalent points"—points in the same position on the object and on the image. After some discussion, they decided that they had to make the two turtles at these equivalent points face each other. To do this they used another tool we had written, named `face`, which output the turn required by one turtle so that it would face another; for example, typing `"blue, face "red` outputted LT 50. Carrie and Susie had also internalised the need for left/right reversal strategy for reflection and immediately deduced that they if they had to turn the blue turtle LT 50 to face the red one, they had to turn the red one RT 50 to face the blue one. They then moved the turtles together using `homein` so the two turtles were now sitting on top of each other facing in opposite directions. Finally, with no discussion, the two girls turned the red turtle RT 90 and drew the mirror; the need for perpendicularity was obvious.

These two examples are indicative of all the approaches used in TM by our sample of students. All were motivated by the metaphor of communicating turtles, turtles which could talk to each other, bump into each other. The turtle had meaning for the students and they were keen to control the turtles so as to draw the mirror. Additionally, all the children could "see" the mirror but before interaction with TM appeared to have little idea how to construct it as the relationships required were neither explicit nor precise. After working together with the TM tools, the students designed (in different ways) a strategy for drawing the mirror line and were able to articulate their method and make it explicit. Their methods were generalisable to all reflections, yet were only meaningful using the tools that we made available in this particular computational environment. Now it could be argued that what they achieved would add little to the observation that a mirror line is where to fold a piece of paper to make corresponding points coincide—it appears to remain locked in the realm of action with little or no apparent awareness of the invariants of reflection. We argue that it offers much more. Embedded in the articulation of the students' descriptions of what to do were notions of symmetry, relationships between lengths and angles which had been experienced through actions, but were also *expressed* through matching images and symbolisation. We conjecture that the students had developed a conceptual framework for these notions; one that would make their appropriation of a more conventional definition following teaching a small and rather natural step.

Which tools the students chose to use meant different relationships were stressed or were ignored, yet whatever path was taken, the students became convinced of the correctness of their method. We suggest that this conviction stems from the balanced mediation of the students' conceptions by the tools of TM; the actions the students were able to perform with their accompanying symbolic representations *connected* with their original visualisations as well as with a mathematical perspective on the problem.

3.2 Using reflection as a problem-solving tool[10]

An important idea in Euclidean constructions is that circles can be used to preserve lengths. We had been working with a group of 12- 13 year-old students from the same school as in the MPB case studies and had noticed that they found this idea very difficult; they thought of circles as objects and were not accustomed to using them as tools in problem solving. This is not altogether surprising given students' experiences of circles in and outside school where the emphasis is on their shape rather than their distinguishing properties. We tried to come up with some simple challenges that would focus attention on using a circle as a means to maintain a constant and equal relationship between two lengths and where the students would come to see the functionality and generality of this "new view". Given the students

[10] Adapted from work with Lulu Healy, Reinhard Hölzl, and Richard Noss.

were working with Cabri[11], we also wanted to achieve our two objectives in a dynamic way. This was our agenda but as the case study shows it did not work out quite like this for the students, and we found them using the notions of reflection and symmetry in a creative way directly suitable for the microworld in which they were exploring.

The children were asked to construct two lines *AB* and *CD* and a point *P* on *AB* as shown in Figure 4.

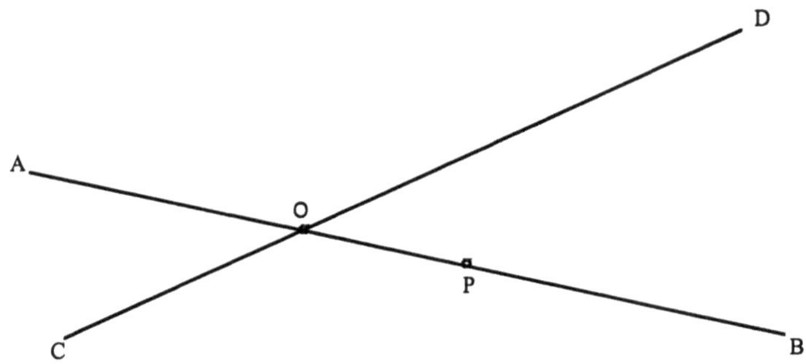

Figure 4: The original screen: Two intersecting lines with a point *P* on one line

Their task was to construct a point *P'* on the other line so that, however *P* was moved, the distance *OP'* remained equal to *OP*. The idea was that they would use a circle to construct this invariant relationship, but as illustrated in the work of two girls, interactions with the available tools that it made sense to them to use, allowed them to come up with a rather surprising solution.

After some experimentation, the girls decided they wanted to use the menu item "symmetrical point", a simple construction operationalised by clicking one point and the "mirror" object. The students had used this before and could predict its outcome. It was though a "black box", and the students were unlikely to be able to articulate exactly what was preserved by clicking. But nonetheless it appeared that this tool had "just the right amount of abstraction built into it"; its outcome was familiar and to some extent predictable, and so by use and reflection on outcome, it was reasonable to assume that the students would be able to find out what it did. In fact it turned out that the meanings they constructed were more flexible than we predicted.

The girls constructed a point, *P'*, the reflection of point *P* in the second line using symmetrical point. Dragging *P* did of course move *P'* in unison—one constraint mastered—but *P'* most certainly did not satisfy the second condition of the task as it did not lie on the second line.

Cleo and Musa started again. They set up the two lines again and proceeded to "solve" the problem by eye: first by simply placing a basic point on *CD* where they

[11] Cabri 1 was used in this case study

thought it should be, and then restricting it to the line by using "point on object". So this time the girls had fixed the second constraint. They now needed to coordinate the two approaches so both task conditions were satisfied. After they had given status to this second point by labelling it *P'*, we asked: "Can it be messed up?" (This term had been coined in the language we had developed with our students to talk about the idea of invariance on dragging, see Healy, Hoelzl, Hoyles & Noss, 1994). It meant "Does *P' always* move with the *P*?" The girls were in no doubt that it did not. In fact, Musa muttered, *"We want P to be a point of intersection"*, indicating that she knew they had to use another object that would intersect with *CD* to "fix" *P'*.

However, the pair were stuck and simply started to guess, adopting a strategy we have frequently observed in computer environments: randomly opening menus and trying out various items in increasing desperation! But the girls kept returning to symmetrical point; they clearly wanted to find a way to use this tool. In fact, the girls repeated their first attempt and constructed again the symmetrical point of *P* in the second line, even though they knew it satisfied only one of the two task constraints. So what now? How could the girls coordinate what they knew—that a symmetrical point in some line would move with *P*—in order to come up with a solution to the task?

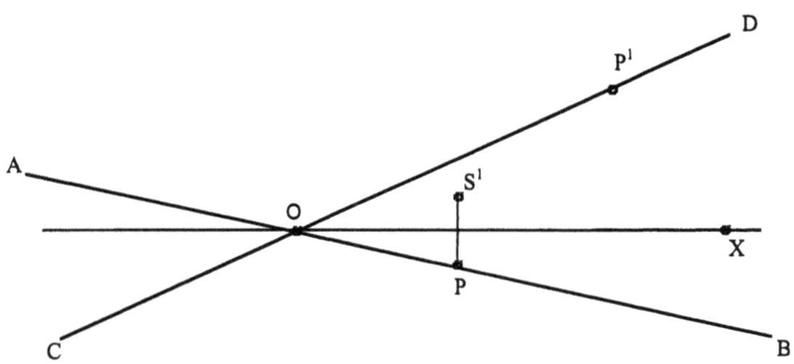

Figure 5: Constructing a symmetrical point in a random line through *O*

They came up with an ingenious solution that exploited the idea of reflection as encapsulated in the tool "symmetrical point". Because of their commitment to the use of this tool their goal had changed to finding a mirror line. It could be described as follows: "Can we find a line in which we can construct a symmetrical point to *P* so that the image lies on *CD*?" Even though they were not as yet able to construct this line, they were clearly convinced that it existed. Their approach could only be experimental, and here again the microworld provided them with just the tool that they needed. They constructed a line segment through *O* to any basic point *X* that could be moved about at random. They then constructed a symmetrical point, *S'*, to *P* in this line *OX* (see Figure 5) and then dragged *X* to move *OX* until *S'* lay on the line *CD* (see Figure 6). They had co-constructed a solution with the computer!

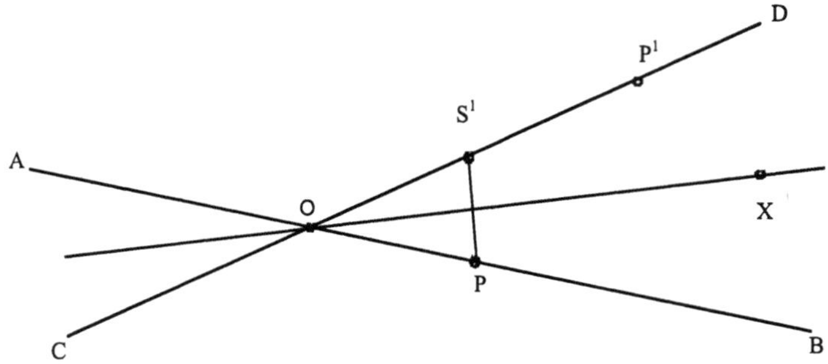

Figure 6: Adjusting *OX* so the symmetrical point lies on *CD*

The girls had used the line, *OX*, as a tool to give meaning to their solution but of course this line had yet to be constructed. But the important point in the building of their solution was that the line was not simply "virtual", but one they could construct, move and think about. They brainstormed and discussed what this line could be, what relationships it must satisfy. Eventually Cleo said, *"It makes two triangles which are the same"*. Musa responded, *"Although the two lines are the same, what else?"* They then came up with the idea that the line *OX* made equal angles with *OP* and *OS'*—that is, it bisected the angle between the two original axes.

For these two girls the task was not finished, because they were unfamiliar with the word *bisector,* as it is called in the Cabri menu (even the word *angle bisector* is not a term used in the classroom). But they (and we) were convinced that the work had been done. They had established a relationship that the line had to satisfy and had simply to find the tool to do this.

There are many interesting facets to this episode. First, note how quickly on their own initiative the girls learnt to label points. They were motivated to communicate with each other (and to us) and it was labelling that provided them with a language to describe their actions and to talk about their pointing and clicking—it was a crucially important communicative tool. Second, it was very apparent that the girls used visual means as an experimental device—they needed to see where to go—to build "something", to flag where it should be and then work out later the relationships it must satisfy (see Noss and Hoyles, 1996). The tools they used, most notably symmetric point as a means to reflect, served an important role in giving meaning to the constructions and focusing attention on relevant relationships. Reflection was not here the object of the task but a tool to solve it. But note too that by choosing to use this tool, the girls had circumvented our agenda. They did not use a circle to construct the required point but rather found a way that exploited something more familiar. This serves as another illustration of how design must take account of tasks and tools but also of student preferences and goals. Tools in the control of the students will be used in unpredictable ways, but their use opens windows on to student meanings and how they co-evolve with the technology.

4. CONCLUSIONS

The case studies have served to illustrate how students solved some geometry problems by exploiting the tools in two different microworlds. In both cases, the microworld served as a modelling tool kit but also as a "concrete resource" by which the student constructions could be shared, manipulated and discussed. In both scenarios, the students were able to articulate their methods and the relationships they used in terms of the medium of construction—they breathed life into the web of resources available and built new meanings from their actions in the microworld. The role of co-construction must not be underestimated; both case studies involved pairs of students participating in the activities and the outcome and the processes by which it was achieved evolved from their joint activity. The students shared their images and together tried to find ways to express them.

I have attempted to illustrate how meanings emerged by students coordinating their own knowledge and understandings through action within the microworlds, building new entities or using tools in new ways, and simultaneously articulating and meshing fragments of knowledge as encapsulated in the computational objects and their relationships. The microworlds were not simply constitutive of the students' behaviour, but were themselves changing as the students created new meanings by interaction within them; in the first case by the building of a new tool and in the second by the use of a familiar tool in a new way. The microworld tools were designed so that a user would develop an appreciation of the (situated) form of generalised relations. This is a key insight from these case studies: parts of a model were built into the fabric of the medium, shaping the types of actions made possible. The level of what could be thought about, talked about, was notched up a rung or two; students constructed situated abstractions of mathematical relations by interacting within the existing model, through "statements" (mouse clicks to produce a symmetrical point, pieces of programs that commanded turtles) which were already expressions of mathematical abstractions. It is the web of relationships and objects offered by the computer that acted both as a support for developing new meanings and as a means for transcending that support. For by manipulating objects and articulating the relationships between them, a dual action/notational framework was developed which became a new resource for learning.

A Cabri bisector or a small procedure like `homein` are themselves abstractions. Both wrapped up much of the process of construction into a set of actions, which the girls in our case studies came to appreciate. The computer "knows" what kinds of objects it "needs" for a bisector or for a program—so some of the mathematical essence comes ready-made; it is a tool to construct objects and at the same time an object which encapsulates a relationship. This duality is at the root of mathematical activity and it is precisely this duality that is so often problematic. Yet actions in microworlds are simultaneously on objects and relationships. As I illustrated in the first case study, tools are not passive: in a microworld the designer's intentions are constituted in the software tools and as we have seen they wrap up some of the mathematical ontology of the environment and form part of the web of ideas and actions embedded in it. Yet it is students who shape these ideas and the functionality and semantics of the invariants planted in the tools—their meanings—shift in the activity. New tools can then be created so that students can step on to a structure with which they empathise. The software supports some actions and not others—but it leaves the critical step of seeing the general in the particular, the theoretical object in its construction, in the hands of the learner.

Learning mathematics does not of course end with activity as illustrated in these case studies. Considering abstraction as situated immediately raises the question of connections between constructions with different tools. As Crook (1994) has suggested from a socio-cultural perspective: "the price to be paid for doubting the notion of generalised thinking skills is that 'some other basis is required for explaining how learners manage to transfer their knowledge from one situation to another" (p. 107). In mathematics, because of its epistemological nature, we need this basis still more and in particular need to forge links to "official" mathematics.

How can the evident gaps be bridged? Construction implies an explicit appreciation of the relationships that have to be respected within any situation, a mathematical model of the situation (how else do you know what to focus on, and what to ignore?). Is it possible for constructions to somehow transcend the medium? We see the generality beyond the tool use, but why should students? Why should they build connections outside the microworld? To make connections between settings means to become aware explicitly of the relationships wrapped into the setting, to notice precisely those elements of the computational web that are interacting with one's current state of understanding. If we consider webbing as a dynamic version of scaffolding, this in the webbing metaphor is the analogue of "fading": not the removal of the support system in the form of the resources of the setting, not the progressive suppression of contextual "props" to "reveal" the mathematical knowledge required for problem solution; but rather its connection with other settings, other notational systems, and other meanings.

But if we consider the question of synthesis across microworlds, we have to ask what remains the same? What is different? Consider the question of constructing a reflection. Paper and pencil construction would demand compasses and it is clear that these actions are only "the same" as the turtle construction or a Cabri construction from a very particular vantage point. At the same time, while we might want to consider a mathematical "essence" shared by, say, the expression of a relationship in Cabri and in geometry, they are certainly not the same for the child, perhaps not the same at all?

Our task is to find ways to assist in the process of forging of connections. How this might be done is a still a matter for research but the differences—and therefore, the difficulties—should not blind us to the potential benefits in taking up the challenge. Although the tools mediate exchanges rather than dictating them, this mediation in a computer microworld can serve to orient the child towards a mathematical viewpoint, because, as I have tried to illustrate, they have to become more aware of the mathematical structure of the tools. The web of mathematical ideas—like its namesake the World Wide Web—may be too complex to understand globally, but local connections are relatively accessible: one way—perhaps the only way—to gain an overview of the Web is to develop for oneself a local collection of familiar connections, and build from there outwards along lines of one's own interests, strengths and experiences.

I have two hypotheses for this proposed research programme: first, there will be no general formula for supporting the "right" agents to give meaning; and second, we must be prepared to consider that in the process of connection both learners and tools will develop new meanings. This is another implication of adopting a socio-cultural approach that applauds the notion of multiple ways of representing reality in approaching any problem, or what is termed "multivoicedness" (Wersch, p. 13). It calls into question any privileging of a particular voice. In mathematics education, do we always have to end up with paper and pencil, static mathematics? One outcome of this research agenda would be to deepen understanding of how far students should notice the mediating role of the tools they use while they explore in mathematical microworlds, and at the same time how the knowledge developed in one setting is "transformed" when activity boundaries are crossed (see Beech, 1999, for a theoretical outline of transfer from the perspective of Activity theory).

In both case studies reported here, pragmatic solutions emerged at the same time as expressions of invariant structures were evoked, rather than preceding them. A fundamental point needs a final reiteration. The resources of the microworld were designed with mathematical intentions in mind, but the students were given the space to share and build their own ideas—to bring them to life, so new tools could emerge in the process of activity This interplay between learners' actions, design and the mathematics "embedded" into a medium is crucial to the microworld story—it is the first step in bridging the gap in traditional mathematical pedagogy between action and expression. The next step is to extend the boundaries of "legitimate" mathematics and to find ways to include the worlds of the particular, the concrete and the manipulable yet which contain within them the seeds of the general and the virtual.

REFERENCES

Beach, K. D. (1999). Consequential transitions: A sociocultural expedition beyond transfer in education. *Review of Research in Education, 24*, 101–139.

Confrey, J., Smith, E., Piliero, S. & Rizzuti, J. (1991). *The use of contextual problems and multi-representational software to teach the concept of functions.* Ithaca, NY (USA), Cornell University, Final project report.

Crook, C. (1994). *Computers and the collaborative experience of learning.* London: Routledge.

Department for Education. (1995). *Mathematics in the National Curriculum.* London: HMSO.

Edwards, L. D. (1992). A Logo microworld for transformation geometry. In C. Hoyles & R. Noss (Eds.), *Learning mathematics and Logo* (pp. 127–155). Cambridge (USA): MIT Press.

Forman, E. A. & Kraker, M. J. (1985). The social origins of logic: The contributions of Piaget and Vygotsky. In M. W. Berkowitz (Ed.), *Peer conflict and psychological growth. New Directions of Child Development, 29*, 23–39.

Freudenthal, H. (1973). *Mathematics as an educational task.* Dordrecht: Reidel.

Grenier, D. (1987). *A teaching process about the notion of reflection for 11 year-old students.* In J. Bergeron, N. Herscovics & C. Kieran (Eds.), Proceedings of the 11th International Conference for the Psychology ofMathematics Education Vol 3, (pp. 124–130) Montreal: University of Quebec & Montreal.

Harel, I. & Papert, S. (Eds.) (1991). *Constructionism.* Norwood (USA): Ablex.

Healy, L. (2002). *Iterative design and comparison of learning systems for reflection in two dimensions.* London (UK), University of London, Ph.D. dissertation.

Healy, L., Hoelzl, R., Hoyles, C. & Noss, R. (1994). Messing up: Reflections on introducing Cabri Géomètre. *Micromath, 10*(1), 14–16.

Healy, L. Pozzi, S. & Hoyles C, (1995). Making sense of groups, computers and mathematics. *Cognition and Instruction, 13*(4), 505–523.

Hoyles,C. (2001) From describing to designing mathematical activity: the next step in developing a social approach to research in mathematics education. In C. Kieran, E. Forman & A. Sfard (Eds.), Special Issue: Bridging the individual and the social: Discursive approaches to research in mathematics education. *Educational Studies in Mathematics, 46*(1–3), 273–286.

Hoyles C, Healy, L., & Pozzi, S. (1990). Interdependence and autonomy: Aspects of groupwork with computers. In H. Mandl, E. De Corte, S. N. Bennett & H. F. Friedrich (Eds), *Learning and instruction: European research in an international context* (Vol. 2.2, pp. 239–257). Oxford: Pergamon.

Hoyles, C., & Healy, L. (1996). *Visualisation, computers and learning.* London (UK), University of London, Final report to the Economic and Social Research Council.

Hoyles, C. & Healy, L. (1997). Un micro-monde pour la symétrie axiale : une base de co-construction de concepts mathématiques ? [A microworld for reflective symmetry: Supporting the co-construction of mathematical ideas?] *Sciences et Techniques Educatives, 4*(1), 67–97.

Hoyles, C., Noss, R. & Pozzi, S (2001). Proportional reasoning in nursing practice. *Journal for Research in Mathematics Education, 32*(1), 4–27.

Kaput, J, Noss, R. & Hoyles, C. (2002). Developing new notations for a learnable mathematics in the computation era. In L. English (Ed.), *Handbook of international research in mathematics education* (pp. 51–75). Mahwah (USA): Lawrence Erlbaum.

Kruger, A. C. (1993). Peer collaboration: Conflict, cooperation or both? In C. Howe (Ed.), *Peer interaction and knowledge acquisition* (pp. 165–182). Oxford: Blackwell.

Küchemann, D.E. (1981). Reflections and rotations. In K. M. Hart (Ed.), *Children's understanding of mathematics, 11–16* (pp. 137–157). London: John Murray.

Lave, J. & Wenger, E. (1991). *Situated learning: Legitimate peripheral participation.* Cambridge: Cambridge University Press

Noss, R. & Hoyles, C. (1996). *Windows on mathematical meanings: Learning cultures and computers.* Dordrecht: Kluwer.

Noss, R.,Hoyles, C. & Pozzi, S. (2002). Abstraction in expertise: A study of nurses' conception of concentration. *Journal for Research in Mathematics Education, 33*(3), 204–229.

Roschelle, J. (1992). Learning by collaboration: Convergent conceptual change. *Journal of the Learning Sciences, 2*(3), 235–276.

Smith, J. P., diSessa, A. A. & Roschelle, J. (1993). Misconceptions reconceived: A constructivist analysis of knowledge in transition. *Journal of the Learning Sciences, 3*(2), 115–163.

Vygotsky, L. S. (1986). *Thought and language.* Cambridge (USA): MIT Press.

Wersch, J. V. (1991). *Voices of the mind: A sociocultural approach to mediated action.* London: Harvester Wheatsheaf.

Wilensky, U. (1993). *Connected mathematics: Building concrete relationships with mathematical knowledge.* Boston (USA), Media Laboratory, Massachusetts Institute of Technology, Ph.D. dissertation.

COLETTE LABORDE

THE HIDDEN ROLE OF DIAGRAMS IN STUDENTS' CONSTRUCTION OF MEANING IN GEOMETRY

This chapter focuses on the use of diagrams at the point when students are beginning to be taught geometry as a coherent field of objects and relations of a theoretical nature. It investigates the relations between the domain of diagrams in paper-and-pencil or software environments and the domain of theoretical objects of geometry, by means of an analysis of students' solution processes when faced with a geometrical task. It is divided into two parts: the first deals with the rules, (sometimes implicit) that govern the use of diagrams in solving school geometry problems; the second describes the actual processes of students in a problem-solving situation, mainly with dynamic geometry software.

1. TWO KINDS OF PROPERTIES

Diagrams in two-dimensional geometry play an ambiguous role: on the one hand, they refer to theoretical geometrical properties, while on the other, they offer *spatio-graphical* properties that can give rise to a student's perceptual activity. Students often conclude that it is possible to construct a geometrical diagram using only visual cues, or to deduce a property empirically by checking the diagram. When students are asked by a teacher to construct a diagram, the teacher expects them to work at the level of geometry using theoretical knowledge, whereas students very often stay at a graphical level and try only to satisfy the visual constraints. For example, the task of drawing a tangent to a circle passing through a given point is frequently viewed by students as the physical task of rotating a straight edge passing through the given point and adjusting it in order to "touch" the circle (Figure 1).

The teacher, on the other hand, is expecting a drawing process based on geometrical relations—the tangent line is perpendicular to the radius, and the locus of points from which it is possible to see a segment under a right angle is a circle. The problem is that the final result in this latter case may not be better from a visual point of view than in the former, with the result that traditional construction problems may fail to call for geometrical knowledge. In these circumstances, rather than helping students, diagrams become an obstacle to geometrical thinking in the sense that they allow students to avoid reasoning in theoretical terms (Fishbein, 1993; Mariotti, 1995; Salin & Berthelot, 1994; Duval, 1995). This leads to the question of the difference between what we call spatio-graphical and theoretical properties in geometry.

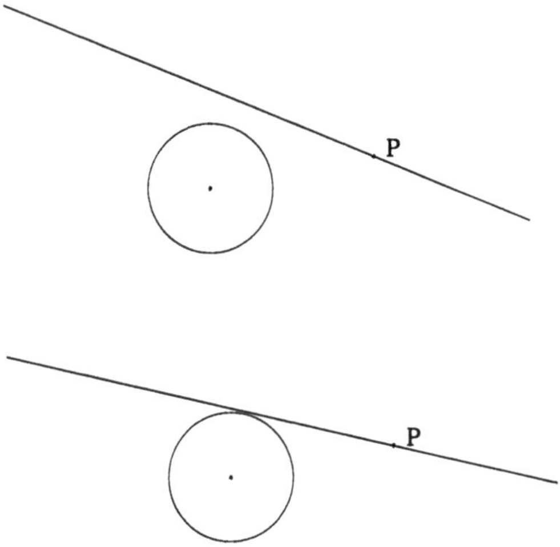

Figure 1

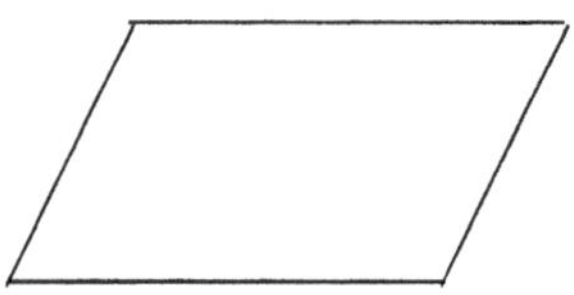

Figure 2

For example, the diagram in Figure 2 represents a parallelogram. It shows several spatio-graphical properties: two sides are horizontal; the other two are oblique in a given direction (bottom left to top right); the opposite sides are parallel; the horizontal sides have a given length. Note that these properties are selected from a larger set of properties like colour or the width of the sides. Some of these spatio-graphical properties can be interpreted in a geometrical way, while others would not be considered interesting from a geometrical point of view: for example, the position of the diagram on the sheet of paper is generally considered to be irrelevant in geometry, as is the slope of the side since it depends on the problem in which the parallelogram occurs. So some spatio-graphical properties of the diagram are *incidental* to the geometrical problem, while others are *necessary* like the parallelism properties. Further, spatio-graphical properties necessarily follow from others: there is a *necessary* link between the parallelism of opposite sides and the fact that the intersecting point of the diagonals is also their midpoint. The teaching of geometry deals with these necessary links between spatio-graphical properties, but one can understand the nature of these links if and only if one also can understand that some other links are merely incidental. *Necessity makes sense in opposition to contingence.* Geometry may appear useful if it allows one to predict, to produce or to explain spatio-graphical properties of diagrams because of these necessary links; but it first requires an awareness of the distinction between such properties and those that are theoretical.

2. THE SHARED ROLE OF DIAGRAMS AND GEOMETRY IN SCHOOL GEOMETRY PROBLEMS

2.1 Different kinds of geometry problems

We start from the distinction between the domain of geometrical objects and relations (denoted by T, referring to *theoretical*) and that of spatio-graphical entities (denoted by SG) instantiated by diagrams on paper, on a computer screen, or by movement produced by a linkage point of a machine. This distinction is very close to that of *conceptual/figural*, made by Fishbein (1993) and Mariotti (1995, pp.100–104). My colleagues and I do not use their terminology, since for Fishbein and Mariotti conceptual and figural refer to two kinds of mental processes of the individual intertwined in a dialectical way. Our terminology is more concerned with referents rather than the mental images and processes of an individual: T denotes the theoretical referents in a geometrical theory, to theoretical objects, relations and operations on these objects as well as to judgments about them that can be expressed in various languages; SG denotes the graphical entities on which it is possible to perform physical actions, and about which it is possible to express ideas, interpretations, opinions, judgments. Straesser (1995, pp. 246–248) also pointed to the distinction between *geometrical* and *graphical representations*, by showing how societal use of graphical representations is frequently pragmatic and only refers to an underlying theory when difficulties are met.

This distinction between T and SG enables a first rough classification of the problems in geometry and in geometry teaching to be made into four kinds of problems:

a) problems internal to either T or SG where the statement of the problem and its solution are both expressed in terms of the same domain: for example, problems in T concerned with proving geometrical properties or problems in SG of reproducing, enlarging, and modifying diagrams;
b) problems that involve moving between T and SG: for example, the definition of a geometrical object is given in T and an SG representation of the object has to be reproduced; or an SG entity is given which has to be interpreted in T.

What is important to is that school problems require the use of both domains and *several moves* between them. We think that this interplay between T and SG is an essential part of the meaning of geometry (a claim also made by Mariotti, 1995, and by Bartolini Bussi in this book). We also consider that this back-and-forth process between T and SG takes place even for experts in school geometry like teachers. The diagram in Figure 3 illustrates the activity of the problem solver according to this view in the case of a problem that starts and ends in the T domain.

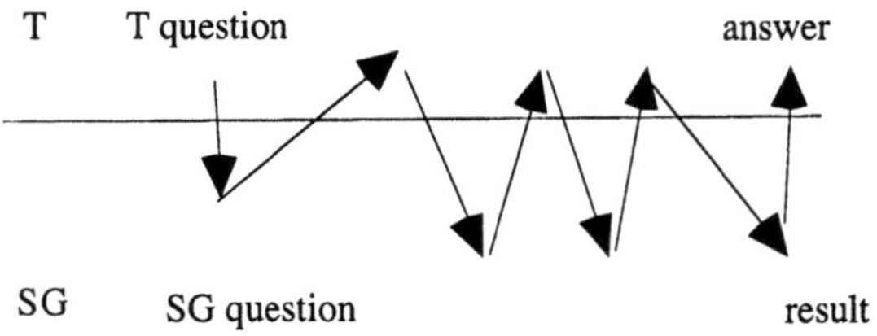

Figure 3

In the same way, an internal SG problem requires several moves between SG and T with the proviso that geometry is an efficient tool in the solution.

2.2 The use of diagrams in the expected solution to school geometry problems

2.2.1 Problems internal to theory

The "official" solution to a T problem must be expressed only in theoretical terms, and the diagram in Figure 4 illustrates the solution expected by a geometry teacher.

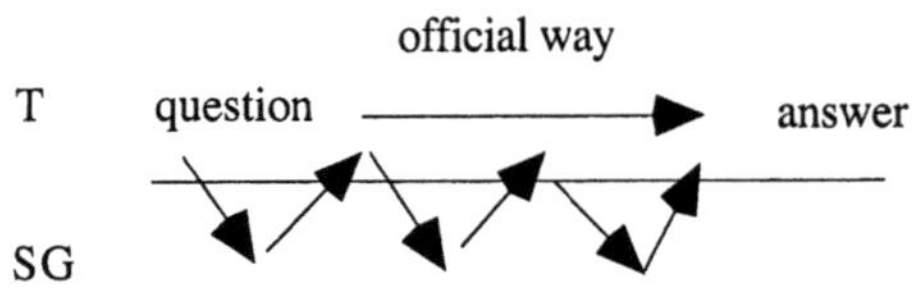

Figure 4

Diagrams are viewed by teachers as auxiliary providers of ideas, but the solution is not allowed to refer to them in any explicit way, as illustrated in the following extract from a French textbook (*Hachette Collège 4ème,* grade 8):

> Some impressions may be wrong. The use of instruments only allows you to get an idea of some properties of a figure [...] When you claim a result, you must prove it by using one or several properties studied in class. [My translation]

But the situation is not so simple. Some information used in proofs is *actually taken from diagrams,* such as all the information related to the *betweenness* of points (in the sense of Hilbert, see Greenberg, 1972), to the orientation of angles (for older students), or to the location of points in regions that are distinguished in a figure. The existence of an intersecting point of lines is also very often (at this school level) taken for granted from the diagram. The proof of the common intersecting point of the three perpendicular bisectors of a triangle given in textbooks (around Grade 8) assumes, without saying it, that the perpendicular bisectors of two sides do in fact intersect. It is taken as obvious, exactly as in Euclid's Elements, where, in the construction of an equilateral triangle with one side *AB* (first proposition of Book 1), the existence of the intersecting points of two circles—the first with centre *A* and radius *AB*, the second with centre *B* and the same radius—was taken for granted. So, we would say that the claim of Hardy (1940) is questionable for mathematics taught at this level:

> It is plain first, that the truth of the theorems which I prove is in no way affected by the quality of my drawings. Their function is merely to bring home my meaning to my hearers, and if I can do that, there would be no gain in having them redrawn by the most skilful draughtsman. (p. 25)

Another example is the official solution expected to a problem for junior or senior high school students. This problem again uses information from a diagram. It deals with a right triangle *ABC* with a right angle at *A* and any point *P* on *BC*. *D* and *E* are the orthogonal projections of *P* onto *AB* and *AC*, respectively, and the problem consists of finding the position of *P* on *BC* where the length of *DE* would be minimal (Figure 5). The most usual solution consists of establishing that *AP* and *DE*, as the diagonals of a rectangle, are equal, and then replacing the minimum of DE by the minimum of *AP*. Then minimising *AP* means minimising the distance from a point to the segment *BC*.

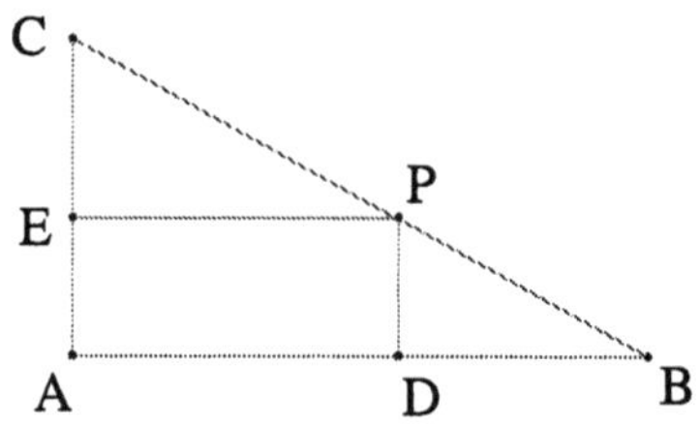

Figure 5

The expected solution claims that this distance is minimal when *P* is the foot of a perpendicular line from *A* to *AC*. The solution thus considers as obvious and does not make explicit that, in this case, the minimal distance to the segment BC is equal to the minimal distance to the straight line BC because the perpendicular line intersects the segment BC. This step in the proof is neither expressed nor justified; it is simply taken for granted from reading the diagram.

These arguments suggest that students must be able not only to distinguish representations from their theoretical referent, but also to know that in some cases they are allowed to use properties of the spatio-graphical representations without justifying them by theoretical arguments—while in other cases they are not allowed to do that.

2.2.2 Problems internal to the spatio-graphical domain

In SG problems where a diagram is to be reproduced identically or enlarged or reduced, the student's task is to take relevant information from the diagram and to use it either directly or in order to deduce the geometrical properties needed to construct the expected diagram. Again, some kinds of information are allowed, while others are not.

The exercise in Figure 6, taken from a French schoolbook for 12- to 13-year-old students (*"Cinq sur cinq" classe de 6ème,* Hachette, 1994, ex. 47, p. 45), illustrates this claim:

Refaire, en plus grand, le dessin suivant [Enlarge the following diagram]:

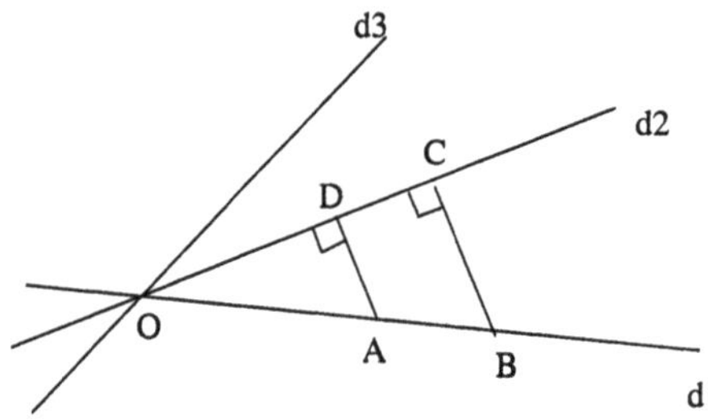

Figure 6

Presumably, the students are not expected to enlarge the given diagram in proportion—signalled by the fact that the length of the segment is not given in whole number of centimetres, and the size of the angles is not given in degrees. It is likely that the students are expected to preserve the following relations and objects:

- three lines d_1, d_2, d_3 intersect in a point O, so that the angles between d_3 and d_2 and between d_2 and d_1 are acute, and d_2 lies between d_1 and d_3;
- two points A and B lie on d_1 in the order O, A, B;
- C and D are orthogonal projections of A and B on d_2.

The two first kinds of objects and relations arise directly from the visualization of the diagram: recognition of straight lines, intersections, points on a line, betweenness and acute angles. The last kind must be inferred from the conventional marks of right angles on the diagram. In reproducing diagrams, it is generally accepted that collinearity of points, intersection of objects, and betweenness will be visually recognised. In contrast, parallelism and perpendicularity cannot be directly assumed and must be inferred from properties indicated on the diagram by conventional marks (equality of angles, for example).

2.2.3 Problems mapping the spatio-graphical and theoretical domains

Construction problems belong to this category. A set of objects and relations is given in T, and the expected product is a diagram in SG but officially produced through knowledge of T.

In the classical problem of this kind coming from antiquity, only a static use of straightedge and compass is allowed. In the Euclidean tradition, movement is forbidden, and rotating a straightedge or transferring a length by an open compass is not allowed. In particular, trial-and-error strategies based on visualisation are not permitted (see the example in Figure 1 of drawing a tangent). But putting points on a line or a circle or producing a point as an intersection is done only visually, which probably explains why collinearity and intersection can be inferred from a diagram.

In conclusion, spatio-graphical and geometrical aspects are intertwined in school geometry, but their respective use by students is controlled by implicit rules partly inherited from the Euclidean tradition and partly from choices emanating from the didactical transposition. We hypothesise that both of these aspects contribute to the meaning ascribed to a geometrical activity, even though the theoretical aspects are mainly stressed in teaching. In the section following the next one, the interrelations of both aspects will be analysed by reference to students' activity while solving geometrical problems.

3. DIAGRAMS IN COMPUTER-BASED ENVIRONMENTS

Spatio-graphical and geometrical aspects are very much interrelated in the new kind of diagrams provided by dynamic geometry environments because their behaviour is controlled by theory. In these environments, like Cabri-géomètre (Laborde & Straesser, 1990) or Geometer's Sketchpad (1993), diagrams result from sequences of primitives expressed in geometrical terms chosen by the user. A crucial feature of these diagrams is their quasi-independence of the user once they have been created: when the user drags one element of the diagram, it is modified according to the geometry of its constructions rather than the wishes of the user. This is not the case in paper-and-pencil diagrams, which can be slightly distorted by the students in order to meet their expectations.

It can be easily assumed that these dynamic geometry environments favour a stronger link between spatio-graphical and geometrical aspects since spatial invariants in the moving diagrams almost certainly represent geometrical invariants.

We attempt to identify in the following section the extent to which this closer link might affect students' solutions.

4. INTERRELATIONS BETWEEN DIAGRAMS AND GEOMETRY IN STUDENTS' SOLUTION PROCESSES

4.1 The case of Cleo and Musa

Let us start with an example from another chapter of this book (see Hoyles). Cleo and Musa were faced with a construction problem on a given Cabri-diagram. Two lines were given, with *P* on one of them. They had to construct a point *P'* on the other line satisfying the constraint, *OP* = *OP'* (Figure 7).

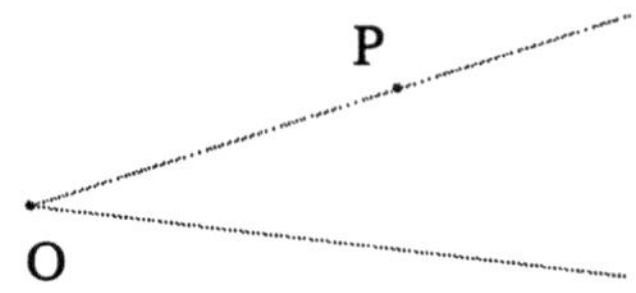

Figure 7

Their solution process was made up of successive moves between the SG and T levels. They started by solving the task on the SG level by placing a point at the right place only by eye—not as a result of any geometric construction. As Hoyles stated, they knew that it was not the solution, in particular because the point obtained in this way would not move with P.

Then they started from their knowledge of reflection (which was not complete) and constructed a reflection of *P* in *OP'*, which obviously did not satisfy the requirement of lying on the line. In this second phase, they linked the T level with the SG level, with the goal of seeing the spatial effect of reflection. This could convince them of the equality *OP* = *OP'*. In the third phase, they could express the problem in geometrical terms in a more precise way: how to find a line in which the symmetrical point of *P* would lie on the other given line? The second phase linking T and SG allowed them better to define the task in T terms. In a fourth step, they systematically explored the effect of a reflection in a variable line *OS* in order to obtain the image of *P* on the other given line. Again, the T task was solved at the SG level. From the visualisation of the position of *OS*, they came up with the equality of the angles *POS* and *SOP'*, which was an interpretation in geometrical terms of the spatial position of *P*.

The successive moves of Cleo and Musa between T and SG well illustrate how the solution is not elaborated straight-away in T and how every step is constructed on the basis of preceding one. The support of the diagram and its dynamical possibilities was also critical in the construction.

The chapter by Bartolini Bussi in this book points to a similar process (that is, a dialectical process between the physical object and the ideal object in the Orthotome task) and stresses how this process is based on visual tactile activity. Bazin (1994) considers that taking information from the diagram must be an essential part of a computer systems expert in solving geometry problems. Pluvinage (1989) and Rauscher (1993) support the view that this ability must be learned at school.

Our aim was to study these moves between T and SG empirically, by giving students geometry tasks and analysing their verbalisations and actions. This approach is close to that of Bartolini Bussi (1991, pp. II-100), who distinguished two types of moves in the solution process of students involved in a geometrical task based on linkages: experimental moves and logical moves. An experimental move is based on visual tactile experiment, while a logical move involves the production of a statement deduced from accepted statements. Because our research group aimed at focusing on the links between diagrams and theory, our method of coding the activity of the students considered, in addition, the moves between what Bussi would call experimental and logical. This method is described in the following section.

4.2 Method of investigation

To investigate these interrelations, we observed students working in pairs on a joint task in geometry with Cabri-géomètre. Their verbal exchanges allowed the researcher to work on the exteriorisation of their approaches and ideas. The choice of the task was crucial in order that the place and the role of the diagram in the problem was not the same in every category distinguished above. Tasks from all the categories were used. Students were audio-taped and their verbal exchanges transcribed; all their actions on the computer (dragging an element of the diagram, use of a menu item, typing on the keyboard, click of the mouse) were also recorded by the computer facility entitled "journal of session".

Once a protocol comprising the transcription of the students' verbalisations and actions and the screen diagrams had been made, it was segmented into units, each of which were ascribed to one of the following three categories:

- referring to spatio-graphical realities, SG;
- referring to geometrical objects and relation, T; and
- establishing a link between T and SG.

The segmentation of the transcription was critical. Working with units that are too small is meaningless. The unit must be at least a proposition in the verbal exchange or an action, so that the whole argumentation or project of the student can be considered. But it may happen that a sentence contains two parts, one of them referring to SG, the other to T. If the sentence linked the two parts, it was considered as matching SG and T. If not, the sentence was split into two units.

Ascribing a category to each unit was by no means easy. We attempted to achieve this on the basis of the words used and the student's actions or gestures. We decided that verbs like *see, make, draw* or movement verbs referred to SG. Although it is perhaps easy to decide that "the triangle crashes" (*le triangle s'écrase*) refers to SG, the sentence "the angle becomes flat" is more ambiguous.

One of the main difficulties in designating a category from words is that although geometry terminology is made up of terms from Latin or Greek (see Howson's chapter in this book), the same terms may be used by students to refer to spatial relations (as Howson stresses, spatial awareness is strongly involved in the meaning of geometrical terms). A reference to theorems or rules and the use of implications were considered as referring to the T domain.

All the tedious work on protocols does, however, offer some advantages: in our case this analysis produced evidence of several phenomena that could well have remained invisible without it. These phenomena include:

- the meaning of the task for each student; without this systematic analysis, we would not have discovered that the task intended to deal with geometry was conceived as a spatio-graphical task in some work phases;
- the various kinds of mapping between SG and T and their circumstances;
- the differences between two students working together, one working at the SG level while the other one at the T level; and
- an unexpected category not planned in our method.

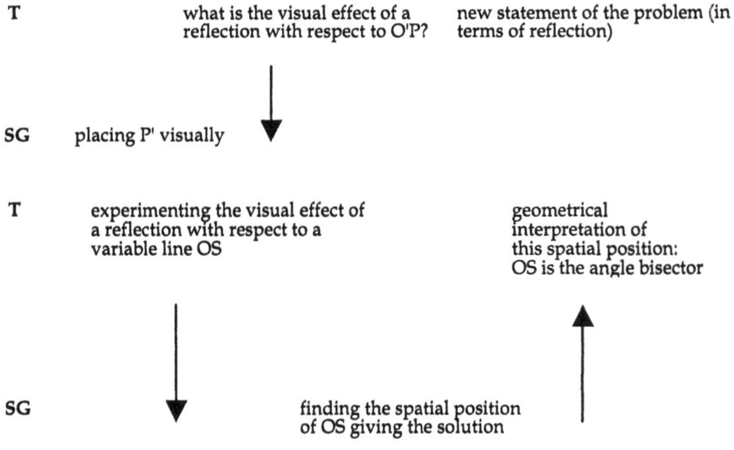

Figure 8

Figure 8 illustrates the categories underlying the problem-solving processes of Cleo and Musa presented earlier. Vertical arrows are directed upward when visual evidence at the SG level is interpreted in geometrical terms; they are directed downward when an action is made at the SG level in order to obtain the spatial representation of a theoretical object on a diagram.

4.3 Meaning of the task

A further complication in our analysis is that a task given by the teacher as a T problem is not necessarily viewed as such by the students, as illustrated below by referring back to the task presented in the introduction to this section (minimising distance *AP* in right angle triangle *ABC*). From the teacher's point of view, this is a task about geometry: the geometrical characterisation of the point of the hypotenuse minimising the length of a segment depending on this point. But the students will always start by drawing the diagram, so the problem is for them at this entry point one in SG: where to put *P* on *BC*? Although students often describe the position of *P* in geometrical terms, it is important to recognise that this behaviour does not necessarily mean that they want to solve a geometry problem, for the following two reasons: first, because of the context of a geometry class; and second, because of Cabri-primitives, which are expressed in geometrical terms. In order to check whether the spatial position of *P* that they have found is correct (i.e., remains invariant in the drag mode), they have to construct it by means of Cabri and therefore have to use Cabri primitives.

The first reason is of a social nature and flowing from what Brousseau (1992) calls the didactical contract: it is part of the obligations of the students in a geometry class to use geometrical terminology. Nonetheless, the problem is not immediately a geometrical problem for students but rather is made up of two phases: finding the spatial position of *P*; followed under the pressure of the didactical contract, by describing this in geometrical terms.

As an illustration, below is a sketch of the evolution of the first half of the work of Paul and Jean-Manuel (14 to15 years old, Grade 9) on the same task as that described in the interlude. It is only at Intervention 125 that Jean-Manuel expresses the necessity of proving—"*Ben oui! Il faut démontrer*" [well, we must prove]—and the first treatment at the geometrical level comes at Intervention 156. All the previous work deals with the Cabri diagram, with some evolution which can be structured into the following phases:

- estimation by eye of the spatial position of *P* (41 to 43);
- determination of the spatial position of *P* by using the measuring facility (44 to 64);
- naming of the position found: it is the midpoint (54);
- experimental checking by using Cabri's primitive midpoint (67 to 77);
- successive trial of some lines, angle bisector then altitude of the triangle and checking whether the point obtained on *BC* coincides with the spatial position found (78 to 112);
- conviction that the foot of the altitude is the right solution (118); and
- dragging the vertices of the triangle *ABC* to be really sure (119 to 122).

In this excerpt, we recognise the conjunction of the geometry class and Cabri effects when the students attempt to define the spatial position of P in geometrical terms. They actually did not solve the problem by using geometry. Rather since they were already convinced of having found the right position, their problem was to find the appropriate terminology. The task was only at the geometry level when they were seeking a proof to satisfy the request for a justification: *justifying* means in the culture of a mathematics class proving by using geometrical reasons. In order to find geometrical reasons, the students analysed the diagram carefully, and the second part of the protocol after Intervention 125 contains more arrows between T and SG levels than earlier. This phenomenon is discussed below.

The students did not search for an explanation as a way to solve the problem of determining P in the T domain, because they could do it otherwise on the Cabri diagram. Thus the task did not contain any intrinsic reason for an explanation through geometry.

4.3.1 A problem of existence

Existence problems are typical of the mathematical problems that belong to theory—that is, problems where the question is to determine whether mathematical objects satisfying a set of conditions actually exist. Only theoretical means can prove the non-existence of objects since it is not possible to provide any pragmatic evidence of them.[1] Based on this assumption, we asked students at Grade 7 to answer the following question: does any triangle exist with two perpendicular angle bisectors?

Analyses of observations of students working with Cabri-géomètre and students in a paper-and pencil-environment were contrasted (Abrougui, 1995). They showed that the tasks did not have the same meaning for the students in the two environments. Very often in the paper-and-pencil environment, students in a desperate struggle to produce the diagram for such a triangle simply drew a great many triangles, thus remaining at the SG level. In the Cabri environment, the task at first was perceived as in SG, but after students observed that the angle bisectors were perpendicular only when the triangle became flat, they started to search for geometrical reasons.

Analysing the task seems to be essential when one is attempting to understand the behaviour of students. In particular, tasks seem not to call directly for the use of theoretical means but are often critically approached at a spatio-graphical level and students move to a geometry level only under other pressure possibly from obligations due to the context of the mathematics class. It also implies that if we want to see students evolve toward the use of geometrical means for other reasons

[1] Several pupils asked whether a triangle with two angle bisectors making an angle of 120 degrees exists, produced an equilateral triangle, measured the angle of the angle bisectors, and wrote "yes because it is just in front of me on the sheet of paper" (or on the screen of the computer). They were aware that the triangle was equilateral and could have checked that the angle of the angle bisectors measured 120° by using the theorem of the sum of angles of a triangle but did not need to do this. The cost of a recourse to theory is higher in this case than the cost of using pragmatic evidence.

than classroom tradition, geometry should bring some economy in contrast to a work on a SG level. It does not imply that even in this latter case they necessarily move to geometry as illustrated by the example of the existence task in the paper and pencil environment.

4.4 Students' links between the spatio-graphical and theoretical domains

From analysis of the protocols, we can distinguish the following categories of links between the two domains:

4.4.1 Linking the SG to the T domain (Arrows from SG level to T level)

Two examples can be given. The first comprises an immediate interpretation of spatial phenomena in geometrical terms: "This is a parallelogram." "The two sides are parallel." "*P* is the intersection of the altitude and the hypotenuse." The interpretation may be direct or through a property observed but not indicated on the diagram. For example, students often called a triangle an isosceles triangle as they observed the equality of the displayed measures of their sides.

My colleagues and I claim that the goal of geometry teaching must be to develop this ability of immediate geometric interpretation by students of spatio-graphical phenomena. The young child pays attention only to appearing/disappearing phenomena or movement against immobility, or may be attracted by closed shapes (Gestalt theory). But phenomena like three points being collinear are not seen by the child. This latter phenomenon is of value only for people having some mathematical background.

The second example in this category is apparent when a geometrical reason is given for something observed in the behaviour of a diagram. This category differs from the preceding one in that the reason is expressed—it is not as immediate as in the latter case or when the student wants to convince his or her partner. What is observed may refer to the fact that two SG phenomena occur simultaneously—there are two simultaneous SG invariants—and the reason for this co-occurrence is expressed theoretically. We consider this as an exteriorisation of the students' awareness of implications of a theoretical nature. Dynamic geometry software may favour this type of recognition by showing in the drag mode the permanence of the conjunction of two varying phenomena. Invariance emerges from variation. To some extent, this type of software may be viewed as providing a reification of the continuity principle of Poncelet (mentioned by Bartolini Bussi in this book and Otte, 1995, p. 4). However some studies show that interpreting the behaviours of a diagram or of elements of a diagram under the drag mode may also be difficult for students. Soury-Lavergne (as cited in Sutherland & Balacheff, 1999) shows very well how the immobility of a point in Cabri-geometry was not related to its geometrical independence of the dragged points. For the student there were two separate worlds, the mechanical world of the computer diagram (in our terms the spatial) and the theoretical (or geometrical).

4.4.2 Linking the T to the SG domain (Arrows from T level to SG level)
An example in this category is a prediction based on geometrical knowledge on what should happen in the spatio-graphical domain, or an experimentation on the diagram also based on geometrical knowledge. Links in this category seem to be favoured by the software facilities of measuring, and especially of dragging where geometrical properties are preserved. We can observe experimentation based on the use of implications at the T level as illustrated below:

> Paul and Jean-Manuel wondered whether *PDAE* was a square or a rectangle. Paul proposed to measure the angle of the diagonals. It turned out to be different from 90°. Paul concluded that it was not a square, because if it were a square, the diagonals should be perpendicular.

The computer environment plays a decisive role in the establishment of these links; it not only enlarges the scope of possible experimentation and of visualisation but also modifies the nature of the feedback. The feedback is visual on the surface but is controlled by the theory underlying the environment.

This modification has two opposing consequences: first, there is no place for doubt, since every spatial behaviour of the diagram may be interpreted as geometrical. Students have only to work on the diagram and do not need to have recourse to geometry as a way of overcoming the doubt. Second, the experiments may become richer and based on more complex geometrical knowledge than in a paper-and-pencil environment.

This tension may be solved in teaching by the choice of task conditions given to students. A problem of existence (see above) seems, for example, to raise more justifications in a Cabri environment than in a paper-and-pencil environment in which students only try to solve by checking in diagrams—at least for students of this age group beginning to learn proof. In a paper-and-pencil environment, students did not try to have recourse to a proof, because they did not solve the problem; whereas in Cabri, they were visually convinced of the non-existence of the triangle and tried to find reasons for this surprising visual phenomenon. As de Villiers (1990) stresses, "Proof is not necessarily a prerequisite for conviction, conviction is far more frequently a prerequisite for proof" (p. 18).

4.5 Asymmetry of students' interpretation of the task

The distinction between the T and SG levels also allowed our research group to become aware of differences between students working together when one was speaking at one level while the other was speaking at another. The following dialogue occurred when Nathalie and Aurélie discovered that the triangle they had drawn became flat when one of its vertices was dragged while trying to obtain two perpendicular angle bisectors (Figure 9).

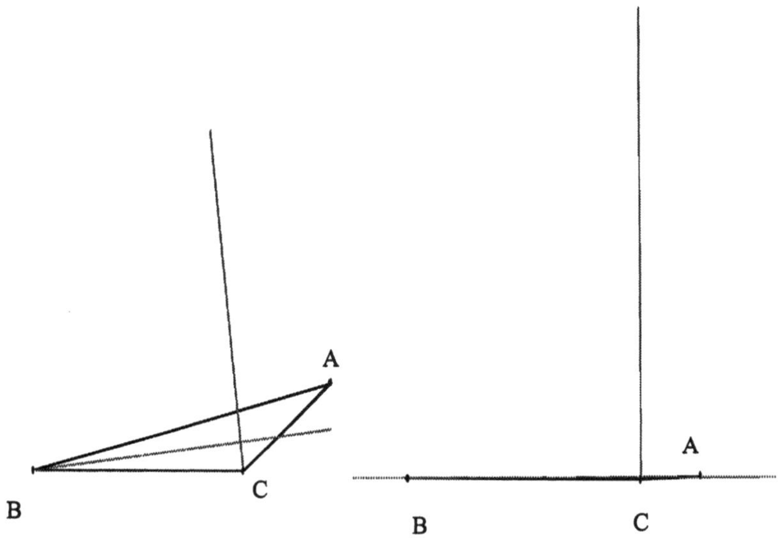

Figure 9

Nathalie asks Aurélie to enlarge the diagram in order to check whether the triangle is really flat. Aurélie describes the spatial situation (it closes the triangle) but starts explaining by using some geometrical knowledge about the sum of angles. Nathalie wants to have better visual evidence and asks Aurélie to use the enlargement facility again. When she said that the triangle is closed, she was obviously referring to what she saw, but Aurélie replied by mentioning that a triangle is always closed (referring here to a topological argument). Nathalie reacted by finding a purely spatial word (the triangle *crashes*), avoiding any ambiguity with mathematics. Aurélie continued by seeking geometrical explanations, but Nathalie remained at the SG level. Even in the final phase of writing their common justification, after Nathalie seemed to have accepted the geometrical justification proposed by Aurélie, Nathalie suggested they write, "*Oui en fait tu mets oui, c'est possible mais à ce moment le triangle ne se voit plus donc disparaît*" [Yes, actually you write down: yes, it is possible, but at this moment the triangle is not visible, therefore it disappears.]. Her answer deals with the spatio-graphical domain: the triangle does not exist as a diagram on the screen.

Aurélie added immediately; "*Ne se voit plus car il devient plat il le devient parce que 90...*" [Is not visible, because it becomes flat; it becomes flat because 90°...]. It is interesting to note that Aurélie handled the transition between SG and T by an intermediate sentence "it becomes flat" that can be interpreted as belonging to SG as well as to T. This led us to come back to our classification and to revise it, the purpose of the next section.

This same asymmetry was evident between Jean-Manuel and Paul. Jean-Manuel discovered by dragging the rectangle *PDAE* (Figure 10) that its diagonals were always equal and found it very exciting ("*c'est génial*").

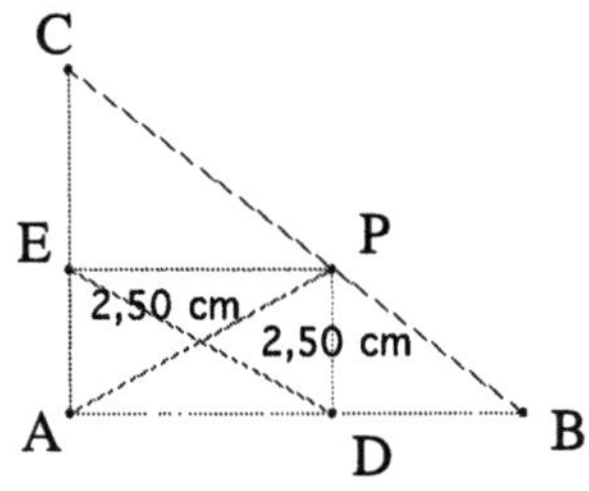

Figure 10

The necessity at the spatio-graphical level of the equality of the two diagonals of a rectangle was really a very interesting discovery for Jean-Manuel as opposed to the random behaviour of these diagonals that he expected. Paul reacted by a laconic: "It is obligatory, you know." Paul's reaction was probably disappointing for Jean-Manuel but was typical of a person acquainted with the fact for a long time. His geometrical knowledge prevented him from enjoying the beauty of the surprise.

4.6 Mixed units

In our systematic coding of protocols, my colleagues and I sometimes could not decide between SG or T for two reasons: the same sentence could be interpreted as referring to SG or T; or the sentence contained terms of both domains strongly mixed. In the first case, these sentences may be considered as expressions allowing a transition between the two levels. A "flat angle" refers both to the spatial image but also, through the term "flat", to a geometry in which the notion of flat angle is related to 180 degrees and supplementary angles; the word scaffolds the move to a geometrical interpretation. It is what Aurélie was doing when she tried to convince Nathalie to give an answer in geometrical terms.

We observed the second category not in our own protocols but in others from abroad. For example, Jones (1998) describes students solving the problem of constructing a circle tangent to two lines and passing through a point *P* lying on one of these lines.

Two English university students, TC and CR, working in Cabri-géomètre began by creating a circle, choosing a centre somewhere between the two lines with *P* as a radial point, and then adjusting its size and position purely by eye so that the circle appeared tangential to the two intersecting lines as in Figure 11.

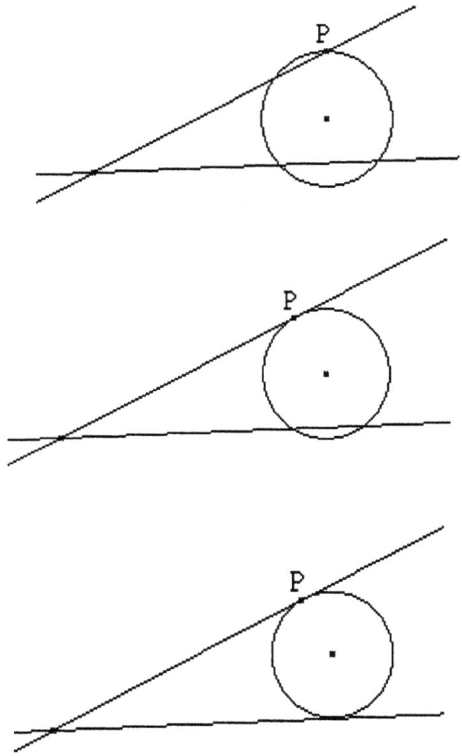

Figure 11

Dragging the centre of the circle led CR to recognise that it lies on the perpendicular at P to the line containing P. They constructed this perpendicular line by means of the software and put a point on this line. They constructed the circle with centre at this point and dragged the centre until they obtained a circle tangent to the second line (Figure 12).

This led CR to recognise that the centre also lies on a perpendicular to the second line and that it is the same distance from the two lines. CR thus came to a mixed determination of the centre, the intersection of a perpendicular line at P and of a moving perpendicular line, ending where the intersection point seems to be the same distance from the lines. The equality of distance mentioned by CR triggered in TC's final step the idea that the centre should be on the angle bisector.

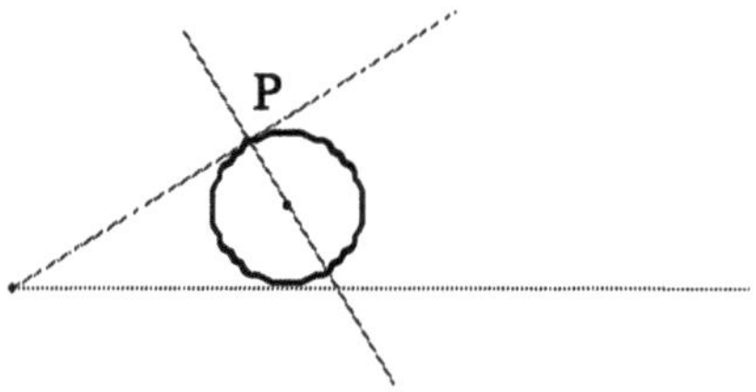

Figure 12

Although CR and TC knew the properties of a circle (equal distance from the centre, perpendicularity of a tangent line to the radius), they did not use them immediately in problem solving but only after recognising the corresponding spatial invariant on the dynamic diagram. These two geometrical properties of the circle were inferred from their spatial representation on the diagram and not recalled from the students' geometrical knowledge. It led to a mixed way of obtaining the centre based on both dynamic spatial and geometrical properties. What is also to be stressed is this mixed determination of the centre was an intermediate step before the final geometrical determination of the centre as belonging to the angle bisector, coming again from the observation of the diagram with the goal of finding a geometrical relation.

TC: There must be a way of securing the centre accurately (p.114)

Examples given in Hoelzl (1994, pp. 151–161; 1995, p. 121) could be interpreted in a similar way. In these examples, it was impossible for us to determine the domain in which the students worked. We would claim that it is may be more fruitful to interpret the work as being at a level that strongly interrelated both domains and as a kind of intermediate state that may evolve toward either geometry or spatio-graphical domain. Interventions by the partner or by the teacher may affect this evolution in a critical way.

The fact that these intermediate steps were found in other protocols (see also Noss & Hoyles, 1996, p. 115) than ours may be due to two reasons: the tasks were construction tasks in which objects satisfying the conjunction of several constraints had to be constructed—so it was more economical to satisfy most in the geometrical domain and the last one in a spatial way. The tasks given in the middle school that we observed were not so complex that the students' strategy of mixing spatial and geometrical means was a way of reducing the complexity. Secondly, the teachers in France may possibly reject these solutions as not being expected—the Cabri construction must remain unchanged by the drag mode. Similarly, the students might consider these solutions as not satisfying the teachers' expectations.

5. CONCLUSION

The distinction between the two domains, the spatio-graphical domain and the geometrical one, allowed us to show how the intertwining of the spatial aspects of diagrams with the theoretical aspects of geometry is especially important at the beginning of learning geometry. However the analysis of school tasks reveals that the kind of information that can be drawn from diagrams and the use to which this information can be put are usually hidden or tacit in teaching, in particular at the point when students begin to be taught about the process of proving. Students must learn these implicit rules for using diagrams in the ways expected by teaching.

The coexistence of spatial and theoretical aspects is thus a source of difficulties and ambiguities. But it is also a source of potential evolution of students in geometry. Learning geometry seems to involve not only learning how to use theoretical statements in deductive reasoning but also learning to recognise visually relevant spatio-graphical invariants attached to geometrical invariants. The analysis of students solving geometry problems in a dynamic geometry environment gave further evidence of back-and-forth moves between the spatio-graphical level and the theoretical level that may play a crucial role. The computer environment acted as a window on students' solving processes and ideas (Noss & Hoyles, 1996). It showed how the first approach of a problem may be purely of spatial nature, and that it may take time before students enter the theoretical domain. It also revealed that after this first approach the activity of students is based on links between both domains. Marrades and Guttierez (2000) also advocate in favour of the idea of a long and slow transition from empirical to formal justifications in a DGS environment. They show how the deductive phase does not appear at the beginning of the solving process but after several empirical approaches and when it appears, how it is related to these empirical approaches.

We also identified actions and verbalisations of students mixing both aspects. We need to know better the role of these mixed statements or actions: do they constitute a necessary step in learning or even in problem-solving activity? To what extent do they contribute to the move to theory? Another point is to formulate conditions for tasks allowing a fruitful interplay between the spatio-graphical and theoretical domains. As reflected by four papers of a special issue devoted to proof in Dynamic Geometry environments (Hadas et al., 2000; Jones, 2000; Marrades & Guttierez, 2000; Mariotti, 2000) it seems that such environments break down the traditional separation between action (as manipulation associated to observation and description) and deduction (as intellectual activity detached from specific objects)and reinforce the moves between the spatial and the theoretical domains.

REFERENCES

Abrougui, H. (1995). *Impact de l'environnement Cabri-géomètre sur les démarches de preuve d'élèves de 5ème dans un problème de construction impossible.* Lyon (France), University of Lyon 1, Mémoire de DEA de Didactique des Disciplines Scientifiques.

Bartolini Bussi, M. (1991). Geometrical proofs and mathematical machines: An exploratory study. In I. Hirabayashi, N. Nohda, K. Shigematsu, F. L. Lin (Eds.), *Proceedings of the XVIIth Conference of the International Group for Psychology of Mathematics Education* (Vol. II, pp. 97–104). Tsukuba (Japan): University of Tsukuba.

Bazin, J.-M. (1994) Géométrie: le rôle de la figure mis en évidence par les difficultés de conception d'un résolveur de problèmes en EIAO. In M. Artigue et al. (Eds.), *Vingt ans de didactique des mathématiques en France* (pp. 371–377). Grenoble: La Pensée Sauvage.

Brousseau, G. (1992). Didactique: What it can do for the teacher. *Recherches en didactique des mathematiques*, Selected papers, 7–40 Grenoble: La Pensée Sauvage.

De Villiers, M. (1990). The role and function of proof in mathematics. *Pythagoras: Journal of the Mathematical Association of Southern Africa, 24,* 17–23.

Duval, R. (1995). Geometrical pictures: Kinds of representation and specific processes. In R. Sutherland & J. Mason (Eds.), *Exploiting mental imagery with computers in mathematics education* (pp. 142–157). Berlin: Springer.

Fishbein, E. (1993). The theory of figural concepts. *Educational Studies in Mathematics, 24,* 139–162.

Geometer's Sketchpad [Computer software from the Visual Geometry Project]. (1993). Berkeley (USA): Key Curriculum Press.

Greenberg, M. (1972). *Euclidean and non Euclidean geometries: Development and history.* NewYork: Freeman.

Hadas N., Hershkowitz R. & Schwarz B. (2000). The role of contradiction and uncertainty in promoting the need to prove in dynamic geometry environments. *Educational Studies in Mathematics, 44*(1–3), 127–150.

Hardy, G. H. (1940). *A mathematician's apology* Cambridge: Cambridge University Press. (Republished 1992.)

Hoelzl, R. (1994). *Im Zugmodus der Cabri-Geometrie.* Weinheim: Deutscher Studien.

Hoelzl, R. (1995). Between drawing and figure. In R. Sutherland & J. Mason (Eds.), *Exploiting mental imagery with computers in mathematics education* (pp. 117–124). Berlin: Springer.

Jones K. (1998) Deductive and intuitive approaches to solving geometrical problems In C. Mammana & V. Villani (Eds.), *Perspectives on the teaching of geometry for the 21st century* (pp.78–83). Dordrecht: Kluwer.

Jones K. (2000). Providing a foundation for deductive reasoning: students' interpretations when using dynamic geometry software and their evolving mathematical explanations. *Educational Studies in Mathematics, 44*(1–3), 55–85.

Laborde, J.-M. & Straesser, R. (1990). Cabri-géomètre: A microworld of geometry for guided discovery learning. *Zentralblatt fuer Didaktik der Mathematik, 5*(90). 171–177.

Marrades R. & Guttierez, A. (2000). Proofs produced by secondary school students learning geometry in a dynamic computer environment. *Educational Studies in Mathematics 44*(1–3), 87–125.

Mariotti, M. A. (1995). Images and concepts in geometrical reasoning. In R. Sutherland & J. Mason (Eds.), *Exploiting mental imagery with computers in mathematics education* (pp. 97–116). Berlin: Springer.

Mariotti M.-A. (2000). Introduction to proof: The mediation of a dynamic software environment. *Educational Studies in Mathematics, 44*(1–3), 25–53.

Noss, R. & Hoyles, C. (1996). *Windows on mathematical meanings: Learning cultures and computers.* Dordrecht: Kluwer.

Otte, M. (1995). Mathematik und Verallgemeinerung. In *Arbeiten aus dem Institut für Didaktik der Mathematik der Universität Bielefeld*, Occasional Paper 158. Bielefeld: University of Bielefeld.

Pluvinage, F. (1989). Aspects multidimensionnels du raisonnement géométrique. *Annales de Didactique et de Sciences Cognitives* (ULP et IREM de Strasbourg), *2*, 5–24.

Rauscher, J. C. (1993). *L'hétérégonéité des professeurs face à des élèves hétérogènes. Le cas de l'enseignement de la géométrie au début du collège.* Strasbourg (France), Université des Sciences Humaines de Strasbourg, Thèse de l'université de Strasbourg.

Salin, M.-H. & Berthelot, R. (1994). Phénomènes liés à l'insertion de situations adidactiques dans l'enseignement élémentaire de la géométrie. In M. Artigue et al. (Eds.), *Vingt ans de didactique des mathématiques en France* (pp.275–282). Grenoble: La Pensée Sauvage.

Straesser, R. (1995). Euclidean versus descriptive: On social needs and teaching geometry. In C. Mammana (Ed.), *Perspectives on the teaching of geometry for the 21st century* (pp. 246–249). Catania (Italy): University of Catania.

Sutherland, R. & Balacheff, N. (1999). Didactical complexity of computational environments for the learning of mathematics. *International Journal of Computers for Mathematical Learning, 4*, 1–26.

JOEL HILLEL AND TOMMY DREYFUS

WHAT'S A BEST FIT? CONSTRUCTION OF MEANING IN A LINEAR ALGEBRA SESSION

Among the various possible senses of "meaning" within mathematics education is the shared meaning constructed by a group of students while dealing with a collection of mathematical tasks and concepts. The Introduction to the section on the Construction of Meaning elaborated on the possible critical components of the communicative acts, which affect the meaning constructed. In this chapter we will try to substantiate how the conditions of communication influence the construction of meaning by examining in detail the transcripts of a linear algebra session of a group of students who were attempting to solve an assigned problem, while working in a Maple lab. The session is replete with different agents of communication including: the students themselves, an observer, the computer and Maple, the classroom teacher (who is not physically present), classroom notes and text. We will examine the changing roles and intentions of these agents, and the formats of communication between them, including different linguistic and symbolic means (diagrams, in particular). We will then analyse how communication supported (or hindered) the processes of solving the assigned task, and the constructing of meaning related to the underlying notions of projection and approximation.

The situation described in this chapter is a rather typical one faced by students of mathematics, namely, an initial loss of meaning, which occurs when a familiar concept is first generalised and then they are asked to reinterpret it in unfamiliar contexts. For example, students may be acquainted with functions as real-valued functions of a real variable given by explicit equations. Their notion of one-to-one function is then usually associated with the "vertical-line test" applied to graphs. When the definition of function is extended to arbitrary sets and students are asked whether a particular function is one-to-one, they have to construct a new meaning for the concept, one that is not tied to the graphical representation. In our case, a more-or-less familiar geometric concept is re-framed within the general theory of vector-spaces. While the generalised notion still retains its geometric referents, students are given a task in which the context is far from the usual geometric one.

The concept involved here is *orthogonal projection*, which, in its geometric setting, is related to the finding the shortest distance from a point to a line (plane) in 2- or 3-dimensional Euclidean space. The fact that the shortest distance is the "perpendicular distance" was part of the solution to the problem cited in the vignette in the introduction to this section. The same situation can be posed in a slightly different but essentially equivalent way: find the vector in a given subspace of R^2 or R^3 which is closest to a given vector **v**, i.e. among all vectors **w** on a line or a plane W, find the one, called **w***, which minimises $||\mathbf{v} - \mathbf{w}||$. The vector **w*** is the "orthogonal projection" of **v** on W since it is orthogonal to the vector **v** – **w*** (see Figure 1).

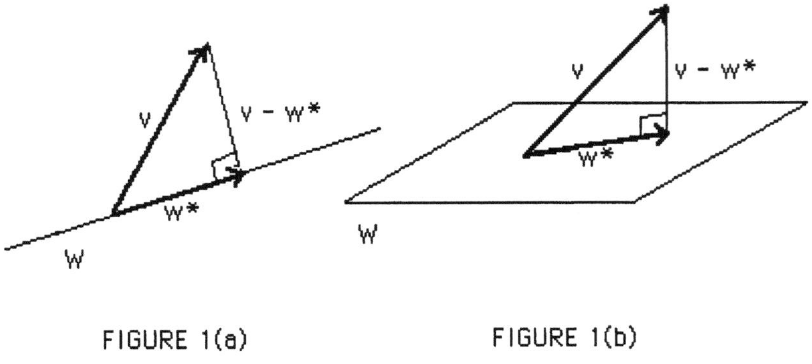

Figure 1

Furthermore, the passage to a coordinate system allows one to express the condition of orthogonality in terms of the dot-product. Thus two vectors in, say, R^3 with coordinates (x_1,x_2,x_3) and (y_1,y_2,y_3) are orthogonal if $x_1y_1 + x_2y_2 + x_3y_3 = 0$. This relation makes it possible to write the coordinates of the orthogonal projection **w*** explicitly in terms of those of the vector **v**.

The problem of finding the vector in a given subspace which is closest to a given vector **v** therefore makes sense in any vector space in which there is an underlying concept of distance and orthogonality. This requires a definition of an "inner-product" of vectors, which is somewhat like the dot-product of vectors in R^n. Thus, the most general setting for defining an orthogonal projection is the so-called *Inner Product Spaces*. While the language used in the most generalised setting remains faithful to the geometric origin, the actual context may be very far from its geometric progenitor. In the session we will be analysing, the students have to deal with a mathematical context involving a vector space of functions defined on a given interval, a subspace of polynomials of degree ≤ 2, and an inner-product of vectors (and hence orthogonality and distance) defined in terms of a definite integral. As we shall see, most of the students' interactions stem from a need to unravel the meaning of several concepts of the general theory in the specific context which is, for them, an unfamiliar terrain.

1. THE TEACHING OF THE TOPIC OF INNER PRODUCT SPACES

We will first describe how the topic of Inner Product Spaces (IPS) fits within a one-year university course in linear algebra. While the description we give is about the practice of one university, even a cursory look at linear algebra textbooks covering the topic of IPS is enough to convince one that this practice corresponds to "the standard routinised presentation of material" (Howson, this volume).

Inner Product Spaces are typically introduced in the second half of a two-semester linear algebra course. Generally, about two weeks of lectures are allotted for the concepts of inner product, orthogonality, length (norm), and projections. Algorithms for obtaining an orthogonal basis for a subspace W from a given basis for W (the Gram-Schmidt method), and finding the projection of a vector **v** on W in terms of the orthogonal basis, are also taught. As for many of the other concepts introduced in the course, the generic example given is that of R^n endowed with the usual dot product <**u**, **v**>:= $x_1y_1 + x_2y_2 + ...x_ny_n$. For this space, students are given standard examples and exercises involving computing norms of vectors, showing that a pair of vectors is or is not orthogonal, applying the Gram-Schmidt procedure, and calculating projections of a vector on a given subspace.

One of the applications of the theory of IPS often included in linear algebra courses is to recast the problem of data-fitting (find the line of best fit through n data points in R^2) as a problem of computing a particular projection in R^n. While this application is accessible and requires little in terms of techniques, it does not serve to justify the generality of the theory. Thus one typically tries to sensitise students to other examples of Inner Product Spaces by introducing examples from two classes of spaces: R^n with a variety of non-standard inner products, and function spaces. The first class seems to serve the purpose of having students go through a ritual—the "practice of techniques in an abstract decontextualised form" (Howson, this volume); no attempts are made to show that such alternative inner products are useful for solving some interesting problems. A typical example from the second class is the space of continuous functions on the interval $[a,b]$, with an inner product defined by the definite integral, i.e. $<f, g>:= \int_a^b f(t)g(t)dt$.

Of course, function spaces are of primary importance in mathematics and mathematical physics and they constitute the real raison-d'être for the general theory of inner product spaces. However, within an introductory linear algebra course such examples are covered very superficially and are often introduced more for their "shock value" than as useful tools. Their purpose is to make students realise that inner product spaces can be quite different from the familiar generic example, and to have them apply the procedure of Gram-Schmidt or of finding a projection in this unfamiliar context. The particular space P_n of polynomial functions of degree $\leq n$ is often used. For example, many texts work out the example of orthogonalising the basis 1, x, x^2 of P_2 on [0,1] with respect to the inner product given by $<p(x),q(x)>:= \int_0^1 p(t)q(t)dt$. It is the problem of projecting the function x^3, defined on [0,1], onto the subspace P_2 that the students dealt in the session described in this chapter.

It thus appears that the basis for choosing and presenting material in standard texts of linear algebra does not differ much from the one used for the school texts described by Howson. As Biehler (this volume) states: "the theory representation [...] is done with regard to certain intended applications inside mathematics". There is some degree of pointing to intended applications within mathematics but this is done in a very superficial manner because it is assumed that students are not ready for these applications.

2. THE SETTING

The students whose work we will be describing here were taking their second one-semester linear algebra course. The whole class was introduced to Maple's Linear Algebra package and part of every assignment contained problems which were to be solved using Maple. The students were expected to go to the computer lab at their own time. The teacher, on occasion, would hand the students a worksheet showing how a problem solved in class is solved with Maple.

A group of six students volunteered, a month after the course had begun, to come once a week to a computer lab in order to work on their assignments. In the lab the students worked mostly in pairs (not necessarily pairing with the same person every time), and they were in sufficiently close proximity to each other that they could interact freely even when not working on the same computer. A different pair of students was audio taped every meeting.

There was an observer (Jay) in the lab who not only helped with technical questions but was also a linear algebra instructor and he occasionally initiated a discussion about the task on hand or about the underlying theory.

The particular session under consideration took place at the end of two lectures devoted to the topic of orthogonality in IPS. Among the examples that were worked out in class, by the students and the teacher, was the derivation of the orthogonal basis $p_0(x) = 1$, $p_1(x) = x-\frac{1}{2}$ and $p_2(x) = x^2-x+\frac{1}{6}$ obtained by applying the Gram-Schmidt procedure to the basis $1, x, x^2$ of $P_2[0,1]$. This orthogonal basis was subsequently used for working out the problem of finding the best quadratic approximation to $\cos(x)$ (In the sequel, we will refer to this as "the $\cos(x)$ task"):

> In the vector space $C[0,1]$ of all continuous functions on [0,1], let $f(x):= \cos(x)$. Find the best approximation of f in the subspace P_2.

Since an orthogonal basis $p_0(x), p_1(x), p_2(x)$ of P_2 was already worked out, the best quadratic approximation of $\cos(x)$ is a quadratic $a_0p_0(x)+a_1p_1(x)+a_2p_2(x)$, and the problem is reduced to evaluating the coefficients a_0, a_1 and a_2, as:

$$a_i = \langle \cos(x), p_i(x)\rangle \Big/ \|p_i(x)\|^2$$

Three of the six students participated in this session. Their assignment was the first one on the topic of IPS, though they may have worked on some problems during the Class Activity time which typically occupied about ten minutes of a 75-minute class. They were provided with a Maple printout of the solution of the $\cos(x)$ task, and were given the following task:

> Let $f(x) := x^3$. Find the best approximation of f in the subspace P_2.

3. AGENTS: INTENTIONS AND CHANGING ROLES

We have identified five active agents in the communicative acts of the session; the human agents, including the students, the teacher, and the observer, and the non-human agents, including Maple, the different mathematical texts (among which a Maple solution of the $\cos(x)$ task that was solved in class, played a major role), and the "canonical diagram" for projection (Figure 1(b))

3.1 Students

The session involved three students. Bernard (B) and Esther (E) were there from the start of the session and were therefore the dominant agents of communication. Tam (T) arrived a bit late and mostly played catch-up. The students, though of "average" mathematical ability, were interested in their work and had already developed an easy and unpretentious style of communication. Their intentions vis-a-vis the task were initially pragmatic, namely, to reach a solution as quickly as possible, and move on to other tasks. Driven by personal and institutional constraints they first adopted a minimalist approach of simply trying to emulate the worked-out example. However, with the progression of the session and the intervention of different agents, the students' intentions became broader and less focused. Among their communicative acts were those that were:

– narrowly focused on getting a solution, e.g.:

> E22: And then we do exactly what she [the teacher] did - I think, I think it's like that.

– attempts to explain their current state of knowledge, e.g.:

> B112: We're given x^3 and we have to approximate it on the plane. We should say why, though.

– indicative of efforts to link with the general theory, e.g.:

E187: But I don't know why we are doing this though. Like what do we have to do, why do we find the product—o.k. here it is, **v**–**w*** ... [Looking through the classroom notes on orthogonality]

3.2 Teacher

The instructor of the course was not present at the session. Nevertheless she communicated with the students through her classroom notes and the sample solution of the $\cos(x)$ task.

In informal conversations with the authors, she indicated that her intention, generally speaking, was to provide students with a good grasp of one of the topics in the linear algebra syllabus, while being cognisant of the range of cognitive difficulties associated with the subject matter. Her classroom notes indicate that she tried to formulate the underlying ideas in an intuitive way. In her reflections on the x^3 task she commented that she didn't expect that the students could "just copy" her solution of the $\cos(x)$ task; that they still needed to sort out what had to change and what can be kept the same (for example, the orthogonal basis of P_2) in going from one example to the other. Furthermore, knowing the particular students involved, she assumed that they would want to gain a better understanding of whatever they were working on.

The students spoke of "her" throughout most of the session and there was no questioning of her authority. The $\cos(x)$ task was not addressed in terms of the general theory (e.g., "this is how the coefficient a_2 is computed"), but rather in terms of "what she did". For example:

E20: Then she did it over there, the integral, so we're going to put the integral of x^3 times this. And then this is going to be the integral of this—it's going to be the exact same thing.

E216: Well, we're doing it just like the way she did it here. She just integrated this and found it between 0 and 1, and then she just divided...

E223: Oh yeah; she gave us a rule with the inner product or something; she wrote it down somewhere.

3.3 Observer

The observer, Jay, was also a linear algebra instructor. His lowest-level role was to help with the technical aspects of using Maple. But he also acted as an observer and a surrogate for the absent class instructor. As an observer, he tried not to intervene and only ask questions in order to clarify the students' actions and understanding (for himself and for them), e.g.:

J63: And what's the connection between least square and inner products?

However, in this particular session, he deliberately attempted to slow down the students' rush to a solution by trying to get them to go beyond the task and to link the particular example to the general theory:

J259: Mhmm. Ok. So you see, I mean it's a very general situation that you have and this is just a particular case of it. What's the vector space you are using in this example? What does it consist of?

At certain junctures during the session, his communication style took on the more traditional teacher-to-student explanations.

The students also seemed to have some expectation that Jay will bail them out if they will run into difficulties:

B23: We'll try it, we'll try it. The teacher should show up anyway soon.

3.4 Computer and Maple

We have analysed the role of Maple in the session in detail elsewhere (Dreyfus & Hillel, 1998). The apparent "intention" of Maple here was to facilitate computations, particularly with regard to definite integrals, and to act as a graphic tool. However, we have noted other roles for Maple. Though it wasn't used at all in the beginning of the session, it was clearly present as a potential partner. The computer was turned on with the Maple prompt "waiting" for an input and the students repeatedly spoke of the need to "tell it" what to do. Maple thus acted as a kind of a "silent moderator"—it facilitated the communication by requiring the students to reach some level of consensus on what "to tell" Maple. Later on, Maple acted as an investigative tool in which students went beyond the given problem by embarking on a "what if" kind of investigation:

B514: So if we would have calculated the integral from 0 to 3 [instead of from 0 to 1] would it have given a good approximation [of the function] from 0 to 3?

3.5 Texts and diagrams

3.5.1 Classroom notes

When the topic of orthogonal projection was introduced, the students were handed three pages of notes, which included the definition of orthogonality, the Gram-

Schmidt method, and a section introducing the idea of "best approximation". The notes communicated the instructor's intention of not merely giving statements of definitions and theorems but to provide the students with some motivation for introducing a concept such as an "orthogonal projection" in the following manner:

> If *V* is an IPS, *W* is a subspace of *V*, and **v** is a vector which does not lie in *W*, then we might be interested in finding the vector in *W* which "best approximates" **v**. This occurs when **v** is a complicated vector and we want to approximate it by something simpler: a subspace is usually something simpler than the whole space. For example, ... in the space of all polynomials you would be approximating a degree 7 polynomial by a polynomial of degree 4. Or, in the space of all continuous functions on a closed interval [0,1], you would be approximating a complicated function by a polynomial of degree 2.

The notes also included two basic theorems; the first stating that the projection of a vector **v** on a subspace *W* is the vector in *W* which is nearest to **v** (the projection of a vector **v** on a subspace *W* was defined as the vector **w*** in *W* with the property that **v**–**w*** is orthogonal to *W*, i.e. for which $\langle \mathbf{v}-\mathbf{w}^*, \mathbf{w}\rangle = 0$ for all **w** in *W*). The second theorem gave the coordinates of **w*** relative to an orthogonal basis for *W*, i.e. $\mathbf{w}^* = a_1\mathbf{u}_1 + \ldots + a_k\mathbf{u}_k$ where $a_j = \langle \mathbf{v},\mathbf{u}_j\rangle/\|\mathbf{u}_j^2\|$. The notes were cross-referenced with the relevant sections from the students' textbook and concluded with some exercises and a homework assignment.

3.5.2 A Maple printout of the cos(x) task

The intention behind the Maple printout of the $\cos(x)$ task was to give the student a nearly prototypical solution, and to remind them of the necessary Maple syntax. It had two particular features, which are worth flagging as they ended up influencing the course of the students' attempts to solve the x^3 task.

First, the coefficient a_0 of $p_0(x)$ was written simply as $\sin(1)$ rather than being evaluated as the quotient

$$\frac{n_0}{d_0} \text{ where } \mathrm{n}_0 := \int_0^1 \cos(x)\cdot 1dx \text{ and } \mathrm{d}_0 := \int_0^1 1\cdot 1dx$$

This kind of simplification-on-the-run is, of course, common practice. When the problem was worked out in class, students were reminded of the full expression for a_0 but in the Maple printout, there was no such reminder. The consequence of such simplifications is that, for the students, the general structure of a particular procedure became hidden.

Second, the printout ended in the plot of the given vector/function $f(x) = \cos(x)$ together with its quadratic projection $q(x)$ (Figure 2). In this example, the two graphs looked fairly close to each other, so the "best" approximation of $f(x)$ appeared indeed to be a very good one. The instructor's intention here was one of good "public relations". While there is no lack of possible examples where the projected vector is actually a very poor fit, such examples are avoided so as not to take the wind out of the students' sails as they grapple with these concepts. Getting a good quadratic fit for cos(x) is a satisfying experience—it gives some power to the theory—a poor fit would undoubtedly leave the students with a "so what" feeling. Furthermore, in this case, one got "for free" an approximating function, which was a good fit on the interval [-2,2] though the best fit was computed only for the relevant interval, [0,1]. This, as we will see, led the students to expect the best fit to always be a good fit.

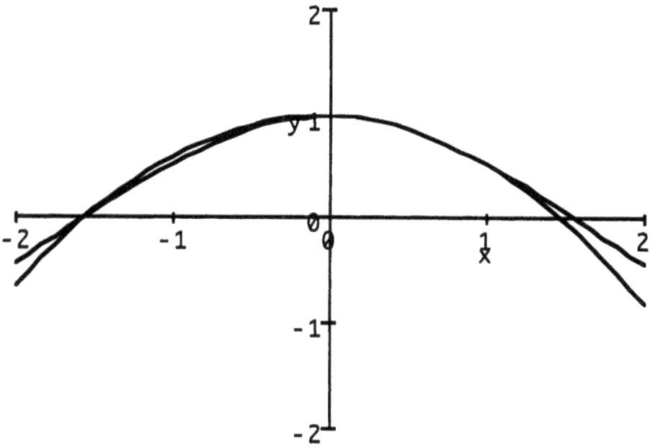

Figure 2: Maple plot of $f(x)$=cos(x) and its best quadratic approximation

3.5.3 Diagram

The general concept of a projection was illustrated, both by the classroom teacher, the textbook, and Jay in the middle of the session, by Figure 1(b) above. The intention of the diagram was to give an intuitive grasp of a projection, based on the familiar situation in 2- and 3-space. It was also meant to be interpreted in a flexible, heuristic way as representing a projection in a general vector space. For the students, however, the diagram became tightly associated with their "concept image" of a projection. It was evoked throughout the session and seemed to be interpreted in a literal way:

E67: The projection of it on that.
B68: On the plane?
E69: Yeah.
B70: Projection on the plane and that's the closest approximation, yeah.

Evoking the image of Figure 1(b) by the students, resulted in a need to clarify the meaning of the diagram and, hence, was one aspect related to their construction of (a local) meaning for a projection.

4. COMMUNICATION AND THE EMERGENCE OF MEANING

In this section, we will examine several episodes of the session in which the students solved the problem of finding the best quadratic approximation of x^3 on the interval [0,1]. We will attempt to show how the different agents and communication acts helped the students in the construction of shared meanings. This construction may be considered as a process of "abstraction in context" (Hershkowitz, Schwarz & Dreyfus, 2001): it had a local character, one which has to do with interpreting some concepts of the theory in a non-familiar context and becoming aware of the different features of the specific context which make it tick.

The episodes are chosen as illustrative of the role of the various agents in affecting shifts in the solution process.

4.1 Episode 1: Two students, model solution

Initially, only Esther and Bernard were present; both had the problem statement and the Maple printout of the solved cos(*x*) example. The computer was turned on, and the Maple prompt appeared on the screen. They quickly recognised that the problem they were asked to solve was to a large extent parallel to the worked out example, so a large proportion of their communication revolved around what they should do rather than where they want to get to or what it means. They seemed to agree, albeit implicitly, on the level of discourse; each contributed bits and pieces which slowly merged into a picture of the procedure to be carried out; the interaction format (in the sense described in the interlude) was of the "interaction between peers" type:

E10: ...I guess we have to do the inner product of just x^3 to find a_0 and then you have to find a_1, and you have to find a_2. Here, I'll show you. See how she did it over here, she put times, it was a_0 plus a_1 times that plus a_2 times that.
B11: Yeah, yeah, yeah, yeah!
E12: So we're going to do the exact same thing, we're going to get something like that.
B13: Yeah but we have to start up here.
E14: Yeah, we have to find our a_0, we find our a_1, and we have to find our a_2. So the way we find a_0 is we just evaluate the derivative of x^3, well the inner product—whatever—we just evaluate it.
B15: The integral!
E16: Yeah , it's going to be uuh the integral...

While the exchange was mostly about procedure, their conversation touched on some terms, which are part of the theory of IPS. For example, right away, Bernard mentioned "orthogonal basis":

B1: Yeah, so we have to find an orthogonal basis first.
E2: Well, we know it already for P_2 'cause we did it in class; 'cause she's asking for P_2.

Esther mentioned "inner product" in E10, though she seemed to confound inner product (which is a bilinear map) with the appearance of an integral sign. This had to do with the "invisibility" of the constant function $p_0(x) = 1$ in the model solution. On the other hand, this same association assured her, on the basis of comparison with the model solution, that she was on the right track, and Bernard clearly confirmed that in B11.

So while their use of mathematical terms was imprecise, it was not wrong, and given the fact that these are new notions, that have been introduced in class for the first time less than a week earlier, the students handled them rather well. This level of imprecision did not disturb the communication between the two students. They seemed to understand each other perfectly well and progressed toward the first step in the actual execution of the task that was set to them. The success of the communication between the students, even on precise mathematical content, appeared to depend less on what one student actually said to the other than on how the other student interpreted the speaker's (possibly vague) statements.

4.2 Episode 2: Three students, model solution

Tam, the third student, joined the session. At first, Tam was simply shown the worked out Maple example and told that they are trying to do the assignment like the example. But Bernard's admonition (B112, see above) that they should tell her why led to a more thorough attempt at explaining the task. Tam, who was playing catch-up, essentially took the role of a "student" asking the other two for clarification. The process of explaining the task to Tam forced Esther and Bernard to be explicit with respect to their understanding of the situation and served to reorganise their current understanding of the problem and the more general context. They managed to make the nature of the task more clear—that x^3 is to be approximated by a second degree polynomial, which, in turn, is written as a linear combination of the polynomials constituting the orthogonal basis of P_2:

E115: Ok. Look, this is the example she gave us.
T116: Ok. Yeah. Ok.
E117: Right? She approximated cos of x.
T118: Mhmm. Ok.
E119: And now we're trying to find an equation like this.
T120: Ok.
E121: 'cause you're doing it with P_2.
T122: Ok.
E123: P_2 is uuh ...
B124: It's an orthogonal basis first of all.
E125: Yeah.
B126: And we're trying to find the a, a_1, a_2.
T127: So here is approximating cos(x) and here, which one do you do?
E128: We're approximating x^3.
T129: Ok. So instead of this one we do x^3.
E130: x^3, yeah.
T131: So, what's uuh...
E132: Well, you have to approximate this to a degree 2 polynomial, right?
T133: Mhmm.

However, the model solution of the cos(x) task was still an active agent so the evaluation of a_0 was explained in a purely instrumental fashion and again referred to incorrectly as the inner product of the initial function:

T142: Why a_0 is sin(1)?
E143: Because the integral of the cos of x is the sin of x...
T144: Oh, ok.
E145: And then she evaluated from 0 to 1, so it's just the sin of 1 and that's what a_0 is...

On the other hand, the "invisibility" of the constant polynomial $p_0(x) = 1$ led Esther to speak of a basis as consisting of only two polynomials.

E134: So this *is* a degree 2 polynomial.
T135: Right.
E136: But this degree 2 polynomial is special because this over here [x – 1/2] and this over here [$x^2 - x + 1/6$] is, uuh, an orthogonal basis.

However, Esther expressed dissatisfaction with the purely instrumental way of finding the coefficients a_i:

E151: Then, after here, we have to find a_1. It's going to be this over this 'cause this is the integral of uuh ... cos of x times this from 0 to 1, and it's going to be this. I don't know why it's that.

Her growing sense that the coefficient a_0 should be part of a pattern governing all coefficients eventually helped her to recover the "missing" polynomial $p_0(x)$ of the orthogonal basis. So when Tam asked a bit later whether the orthogonal basis of P_2 depends on the function being approximated, Esther was fairly complete in her explanation:

T350: You always put 1/2 here [referring to $p_1(x):= x - 1/2$] no matter which one you want to approximate?
E351: It's because they were doing it under P_2.
T352: Yeah
E353: In P_2 we found an orthogonal basis with the "Gram-Shkim" and it is 1, x it's 1/2, and x^2, I think it was +x+1/6 or something. This is the basis we found, minus over here. So that's why we're always using this basis

The communication among the students relating to the activity within a vector space setting was still imprecise. Examples were Bernard's "*The inner product is defined as being the integral from 0 to 1*" (B139) and "*it's an orthogonal basis first of all*" (B124) in reference to the subspace P_2. Esther talked of P_2 as a (geometric) plane and somewhat later of the "*inner product of this* [cos(x)] *times 1*" (E290). Strictly speaking, these are incorrect descriptions, e.g. P_2 is a subspace rather than a basis; it is three dimensional rather than a plane. On the other hand, neither students nor even mathematicians usually express themselves very precisely and correctly when they are in the middle of solving a problem. And the students' statements did indeed express some knowledge of the general structure of the problem setting. The students' understanding of the abstract setting, after this episode, can thus be characterised as budding and still extremely fragile.

4.3 Episode 3: Three students, class notes

This episode had a very different character from the previous one. Here the students tried to sort out the relationship between the underlying theory and the procedures they knew how to carry out. They seemed to be conscious that a deeper meaning in terms of abstract vector space theory is associated with the calculations they were carrying out but, at least for now, they were unable to establish the connection and thus grasp that deeper meaning. For the first time, they abandoned the model solution as their main source of understanding of the problem, and resorted to examining the classroom notes.

E187:	But I *don't* know why we are doing this though. Like what do we have to do, why do we find the product—o.k. here it is, **v**–**w*** ... [Looking through the classroom notes on orthogonality]
B188:	**v**–**w***, **w*** is the one on the plane which is the projection.
E189:	Yeah.
B190:	When **w** is our original vector...
T191:	Yeah, so the dot product is...
E192:	Our original vector is **v**, no? **w** is **v** minus uuh
T193:	**v**
E&B194:	So what's **w**?
T195:	**w*** is the projection of **v**.
E196:	Mhmm, projection of **v** on *W*; the plane *W*.
T197:	Right.
B198:	Mhmm.
E199:	So what's your **w** over here? For every **w** in *W*—oh it doesn't matter!

The students began to draw in some aspects of the general theory. Words such as Gram-Schmidt method, projection and inner product evoked by one student were no longer ignored by the others but provoked reactions (most of which are still not strictly speaking correct); there appeared to be an attempt to slowly make these abstract notions grow into a web of ideas that in the end will somehow congregate into a helpful picture of the theory.

It is interesting that this attempt at theorisation occurred after a detailed description of what needs to be done to compute the coefficients was given by Esther with confirmations and short contributions interjected by Bernard and occasional questions by Tam. However, Esther was not satisfied with such an approach. At several junctures she expresses this by saying *I don't know why we are doing this*. In Howson's terms (this volume), this can be interpreted as a request for "local" meaning for the process they are presently carrying out; it is not a request for meaning in the more global sense of "what's the purpose of all this" or "what's the purpose of learning linear algebra, anyway?" We also note that this episode marked the first time that they spoke about mathematical results without referring to their (absent) instructor—the "she" disappeared and the reference was to the mathematical text.

The students ended up in a state of conflicting intentions. They seemed aware that there was a deeper meaning hidden there, but they also knew that they can probably solve the task they were given without worrying about that meaning.

4.4 Episode 4: Three students, observer-teacher

In this episode, Jay attempted to push the students a little further toward establishing the connection between the procedures they propose to carry out and the theoretical origin of these procedures. He raised the general problem of finding the vector in a subspace U of an IPS V, which is closest to a given **v** in V, and what the coefficients of the projected vector are.

His intervention radically changed the format of interaction that was preponderant up to now into a typical student-teacher interaction format. In an attempt to make the students see the particular example in the light of the general theory, Jay reverted to an expository format, writing on the board and drawing the canonical diagram (Figure 1b) for projection, followed by a series of questions to the students:

J259: Mhmm. Ok. So you see, I mean it's a very general situation that you have and this is just a particular case of it. What's the vector space you are using in this example? What does it consist of?
B260: P_2.
J261: Hmm?
B262: P_2.
J263: Which is the P_2?
B264: Polynomials of degree 2.
J265: Is that your U or your V? You are projecting on a space of polynomials of degree 2.
B266: Of degree 2, yeah.
J267: So that's your...
B268: Our U.
J269: Your U. But what's the vector space you are working with? What kind of vector space is it?
E270: Well, it's all the polynomials—
J271: But if it's the polynomials, where does the cosine come in? I mean... [Jay's explanation omitted]
B272: I don't know.
J273: ...What am I defining the inner product space on, what kind of space?
E274: An inner product space?
J275: Yes, but an inner product space starts with a vector space.
B276: Good try!
J277: What is the vector space that I am starting out with?
E278: P_2, no? 'cause it's orthogonal...

The attempt by Jay to help the students identify the underlying vector space V in the example, generated only irrelevant, short responses, silence, and at the end, nervous laughter. It wasn't really a crucial bit of knowledge for following the steps of the given example, nor for solving the given task. The cos(x) task made explicit mention of the vector space $C[0,1]$ of continuous functions on the interval [0,1], but neither the model solution nor the given problem made the vector-space explicit. The students appeared to fail to see its relevance even after Jay's rather detailed explanation in J271-J273. They had a much better grasp of P_2, which they mentioned repeatedly, but still only in terms of *the plane* P_2 rather than as a subspace of some underlying larger vector space.

4.5 Episode 5: Three students, observer, diagram

This episode occurred after the students had solved the task using Maple to calculate the required coefficients, and ended up with $q(x)$: $= 0.05 - 0.6x + 1.5x^2$ as the best approximation of x^3. They decided (after being urged by Jay) to plot the two functions, and they opted for the interval [-3,3]. Their choice for the interval was not mitigated by any mathematical considerations. The meaning for the underlying space of functions (to the extent that they thought in these terms) was undifferentiated: They saw "functions" rather than "functions on the interval [0,1]" as the primary objects. Thus, they simply noticed that in the cos(x)) task, the domain [-2,2] had been chosen, and they decided to be a bit more bold and to choose [-3,3] as the domain. They obtained the plot in Figure 3.

The plot came as a complete surprise and Bernard couldn't hide his disappointment:

B498: ...but I mean at 2 they are already so far apart!
[...]
B505: But that's supposed to be the best, isn't it?

They obviously expected the fit to be good and, apparently, to be good throughout any interval they pick. The visual feedback by means of the plot led them to discuss the term "best" in "best approximation". They expected a "good" approximation ("good" needs to satisfy their intuition of a function close to the given one), rather than one that is "better than any other". This conception of best fit was reinforced by the model solution, since in that case, the quadratic approximation was not only a very good fit for cos(x) on [0,1], but the fit was good even beyond the relevant interval [0,1]. Jay tried to redirect their attention to the underlying interval:

J506: ...well it's the best over which interval?
B507: P_2.
J508: Hmm?
E509: Oh, yeah, we did this on 0 and 1, the best approximation—'cause we integrated always from 0 to 1, I guess.

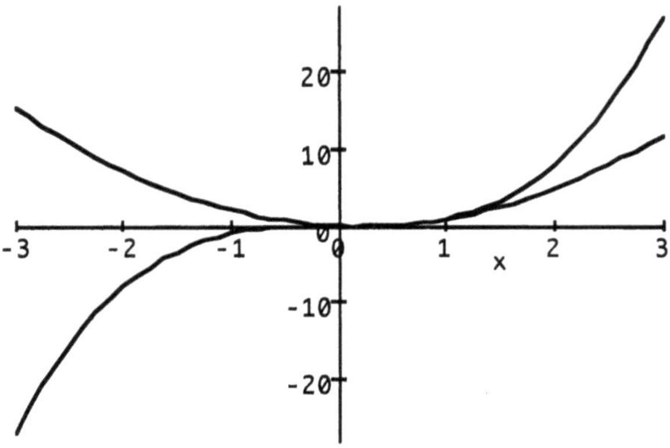

Figure 3: Maple plot of $f(x)=x^3$ and its best quadratic approximation on [0,1]

Esther correctly related the visual feedback to the interval of integration, but her "*I guess*", in E509, indicated that she was not sure what exactly the connection is. Bernard appeared unconvinced altogether, but he came up with a question:

B514: So if we would have calculated the integral from 0 to 3, would it have given a good approximation from 0 to 3?
J515: Well, why don't you try it? Don't ask, that's the nice thing [about Maple]...
[...]
E519: Oh then you want to do everything all over from 0 to 3?
B520: Let's see if it uuh...
E521: Yeah, yeah, yeah!
[...]
B523: ...if its going to give us a good approximation between 0 and 3.

Bernard's proposal is interesting not only because he systematically used "good approximation" rather than "best approximation" but also from the point of view of the responsibility he took for what was happening: He asked a question, expanded the task, an action which up to now was the prerogative of the teacher—there was a change in the intentions, and possibly the status of Bernard as a student. He wanted visual feedback for confirming his newly found meaning rather than a theoretical confirmation. He was likely no more convinced with the theory than Gauss's predecessors had been with that of complex numbers, and like them he wanted confirmation through the senses which, in the case of mathematicians, often means through a geometric representation (Howson, this volume).

The graphical feedback, which forms part of the interaction with Maple, was crucial for the students' learning experience during this episode. They were surprised by the graphical output and found something to be discussed there. The meaning they associated with the term "approximating polynomial" after the discussion was more elaborate than before: it included the idea that the approximation might be "good" on a limited interval only. On the other hand, a plot of $q(x)$: $= 0.05 - 0.6x + 1.5\,x^2$ and x^3 on the interval [0,1], probably reinforced the idea that the approximation is always good on such interval.

4.6 Episode 6: Three students, Maple

It is exactly the kind of "let's see what happens if we change" activities such as the one proposed by Bernard, which are facilitated by the use of Maple. In this particular case, the new problem involved a new inner product which seemed to require making only one simple change: every time the integral from 0 to 1 appeared, 1 had to be replaced by 3. This change was perceived as easy by the students:

E635: Ok. So this would be what, the integral of this—oh just change the 3—ok, this is going to be easy.

No one present in the lab (including Jay) paid attention to the fact that the basis vectors

$$p_0(x) = 1$$

$$p_1(x) = x - \frac{1}{2}$$

$$p_2(x) = x^2 - x + \frac{1}{6}$$

were no longer orthogonal with respect to what essentially is a new inner product. This was easy to overlook because these basis vectors were not computed during the session and they have been referred to constantly as being "the orthogonal basis of P_2" without reference to the interval or the inner product. The resulting graph was unexpected, disappointing and, to some extent, undermined the students' just acquired insights about the relation between a function and its projection over the fundamental interval.

Bernard articulated everyone's astonishment:

B685: Why, what happened?

[...]
B700: Why should [...unclear...] Not even at zero? They should both be zero.
[...]
E704: Yeah, that should be at zero; that's what I'm saying.

To their credit, the students didn't accept the validity of the solution offered by Maple. Their constructed meaning for the approximation at this stage was that it should be a very good approximation if one adheres to the right interval. However, the subtle nature of the difficulty remained elusive for the rest of the session. Only in the following session, a week later, Esther demonstrated a new found understanding of the situation by saying casually "*Oh, we just forgot to find the new orthogonal basis*".

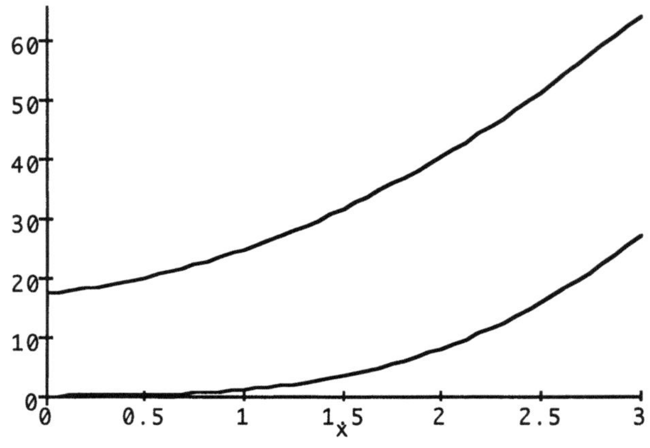

Figure 4: Maple plot of $f(x)=x^3$ and an erroneous quadratic approximation on [0,3]

5. REFLECTIONS ON THE SESSION

5.1 Abstraction in context

The topic of Inner Product Spaces, like most topics taught in linear algebra, coexists in three different universes—the geometrical, the concrete-algebraic (n-tuples with the usual dot-product), and the abstract. IPS and concomitant notions such as orthogonality and projections are usually first introduced in the abstract setting (though the language used retains a geometric flavour), then illustrated geometrically in Euclidean 2- and 3-space, and most of the exercises and examples are set in the concrete-algebraic context. This moving back and forth from one level of description to another is often very confusing for students (Hillel, 2000). Some are only comfortable at the concrete-algebraic level and do not easily accept anything but n-tuples as vectors. Dealing with function spaces requires a certain level of mathematical maturity, which is often lacking in students taking the first linear algebra course.

In the above episodes we have described communication acts and the role of the concomitant agents in the emergence, for a group of students, of mathematical meaning of the general notions of "projection" and "best fit" in the context of a function space. Though we did not attempt to probe into the individual student's understanding of these topics, we can infer from their conversations that a need for reorganising their knowledge arose from the requirements of the task. As the session progressed, the students did construct a deeper meaning of some of the underlying notions by tying together concepts that were, for them, initially disparate. If we look, for example, at the notion of projection, we note that it is initially described vaguely as:

> ...you can do the *image* or something like that!
>
> the *projection* of it on that
>
> projection on the *plane*

But the task imposed the need to think in terms of a new structure, since functions were involved. A bit later in the session, the students spoke in terms of:

> we are *given* x^3 and we have to approximate it on the *plane*
>
> projection of it [x^3 in this case] on the *plane of* P_2

Thus, while the evoked image of Figure 1(b) reinforced the idea that one needed to project "something" on a plane, we see that they have started adding some new levels of generality and structure, so that in the first instance, the "something" was a function rather than a geometric vector, and in the second instance the "plane" itself is P_2 which they have earlier recognised as consisting of polynomial functions. Though they were still mixing up the geometric with the more general aspects of vector spaces, there was an initial attempt here to reorganise some notions into a new structure. Later on in the session they referred again to the "canonical" geometric diagram of a projection which they found in their class notes:

> The projection of **v** on W—the plane W.

Though the reference here was again geometrical, the use of the symbols **v** and W suggested the potential for generality. In fact, late in the session, projection was spoken of in general terms, when Esther declared:

> You know your *vector*, you take your projection—it is the closest one on that *subspace* to that [vector].

We can thus view the students' emerging knowledge as a process of abstraction in context (Dreyfus, Hershkowitz & Hillel, 2001). Though this process did not lead the group of students to completely reorganise their previous notion of vectors so as to include classes of functions, it did contribute to their construction of new meanings for the notions of projection and inner product, through a communicative process involving peers, the teacher, the observer, texts, the computer, Maple and diagrams.

5.2 Students' interactions

Looking across the session, the students' interactions were characterised by several features:

- They made attempts at explaining to each other what they understand; this happened not only while they were in the process of solving the task but also when another student joined them.
- They discussed not only procedural aspects but also theoretical aspects of their work although usually they talked about mathematics in fragmented and very short sentences.
- Each student, in his or her own way, contributed something to facilitate the communication and the ensuing meaning. Esther has a good mathematical sense and was also fairly up to date with the material covered in class to be able to keep the discussion focused and to provide good intuitive explanations. Bernard added a strong social sense, and he pointed out the need to brief the late-comers about what has been going on. He also kept attempting to see the larger picture, particularly at the stage when they were looking for an explanation of what has gone awry. Tam, by virtue of not being up to date, asked questions rather than offered explanations. But this format of interaction spurred attempts by the others to explain and thus also contributed to the collective understanding.

Initially, when Bernard and Esther declared their intention to "do the exact same thing", the model example and, by proxy, the lecturer, were the prominent agents. The computer was turned on but remained virtually untouched. The initial exchanges were characterised by procedural descriptions of what there was to be done interspersed by attempts to pull in relevant terms from the theory. There was plenty, which was vague, imprecise, and even wrong. Yet, somehow, the communication succeeded in moving the students' solution of the task forward. Tam's arrival brought a change to the style of communication. Though she didn't contribute directly to the mathematical discourse, her passive role resulted in attempts to explain to her what the task was about and this led in turn to a reorganisation of the students' knowledge. They were able, at that juncture, to draw more on the implicit theoretical underpinning and the Maple example became less dominant. There was some interaction with the written classroom notes albeit at a very superficial level. The text was not read thoroughly to gain understanding but was rather scanned quickly in order to find something relevant to the problem.

Jay's asked several key questions and even reverted briefly to a lecture style, writing on the board, drawing the "canonical" diagram and taking about projections in a generalised setting. His intervention was only partially successful and his attempt to have the students see the task in its full context (i.e., vector space of functions) did not leave much of an impression. Later on, he encouraged the students to plot the functions they obtained, and to explore the consequences of suggested changes to the task. Both times, these "nudges" brought about Maple action and resulted in some unexpected phenomena, which led to a changed meaning of the problem.

5.3 *The role of diagrams*

Some of the observations during the episodes implicitly or explicitly concern the students' use or non-use of diagrammatic representations. Two types of diagrams played a role during the session. Both are representations of the given function and an approximating polynomial. We have already mentioned one, namely, the canonical diagram of a projection (Figure 1b). The other is a function graph comparing the given function to an approximating polynomial (Figures 2, 3, 4).

These diagrams are quite different and it is not at all obvious how they are linked. There is the matter of interpreting the canonical diagram. On one level, it is as concrete as the function graph for it depicts the projection of a vector in R^3 on a plane. But the canonical diagram is also meant to be general—it is used to symbolise any orthogonal projection in any vector space. It was used as such by the lecturer in the lecture preceding the session and again by Jay when he attempts (in episode 4) to clarify the abstract theoretical framework. Such a diagram is introduced by both teachers and textbook authors for its heuristic value, as a "visual metaphor" (Howson, this volume) for thinking about the notions in the abstract. The difference, however, between the use of this geometric diagram to talk about projection of a vector as opposed to, say, referring to equations in term of balances, is that the object of the metaphor here is itself a special instance of the general theory. Thus it is possible that the students were operating mostly at the "image having level" rather than the "property noticing level" (Pirie and Kieren, 1994) when they evoked the image of the diagram. That is, they focused on the object of the metaphor ("projection in an Inner Product Space *is*...") rather than focus on the properties ("projection in an Inner Product Space *is like*...").

The second type of diagram is specific to function spaces and, in this sense, much more concrete. Reasoning based on function graphs was paramount in the students' best fit considerations, so much so that no other than a graphical definition of best fit is ever used. (In fact, it is quite likely that the students' notion of fit was related to that of the area between the curves rather than the least-square distance.) The students were clearly able to interpret the function graphs very well; they understood the plot sufficiently well to realise what it means, to be surprised at the lack of fit, and to feel the need to overcome that surprise by finding an explanation for the lack of fit. They were also able to control and manipulate the graphical representation, which they clearly couldn't do with the geometric representation. However, their thinking with and about the graphical representation appeared to be affected little, if at all, by vector space considerations.

The way the students dealt with both diagrams suggests some conditions for the efficiency of diagrams as reasoning tools:

1. The students are well acquainted with function graphs and know how to interpret them (though they, perhaps, do not pay much attention to a restriction of the domain of a function such as was the case with the given task where the functions were restricted to the interval [0,1]). On the other hand, they do not know how to interpret the canonical diagram, except in R^3; we have no

indications that they understand the symbolic nature of the elements of that diagram. Clearly, diagrams can only serve as useful reasoning tools, if their elements and the relationships between them are meaningful to the students. The association of such meaning with diagrams is not automatic but needs specific attention in the teaching-learning process. For most types of diagrams such attention is rarely given; the exception are function graphs, mainly because student misinterpretations of such graphs have obtained some publicity for close to twenty years (see, e. g., Janvier, 1978; Schoenfeld, Smith & Arcavi, 1993).

2. In this session, the function graphs come as feedback from Maple; and they are surprising. Surprising feedback is often an incentive for students to feel the need for explanation and even proof (Dreyfus & Hadas, 1996).

6. CONCLUSION

As emphasised by Sierpinska (this volume), any study of communication within mathematics education must take into account the mathematical contents of the communication but also the "mathematical meanings as they are constructed in the direct or mediated interactions between humans" and, we may add, non-human agents as well. A mathematically sophisticated person may look at the content in these terms: there is a general theory (of IPS), there is a specific instance of such spaces (function spaces with a particular inner product) and there is a particular example of a vector and its projection in the specific space. Leaving aside the role of the theory of IPS within mathematics, one can view the situation on hand simply as a case of translating from the general to the specific. This is a simplistic, top-down view of a person who is familiar with the general theory. On the other hand, when we examine the meaning constructed by the students, we see no evidence of a unidirectional flow from the general to the particular. Rather, the general and the specific context as well as the example and the task were all on equal footing, understood, initially, in fragmented and vague terms which became less vague and more coherent as the session progressed.

The socially constructed meaning was mediated by different agents. Some of the agents were deliberately put in place in order to enhance the process—students were encouraged to work collectively and in a computer environment. Other agents (text, notebooks, teacher) were always implicitly present in such activities. We have tried to glean from the transcripts of the students' communication how the agents' roles and contribution to the construction of meaning evolved during the session. In some cases, the contributions were quite explicit (e.g., Maple's feedback of a poor fit between the function and its projection), in others, less obvious and more conjectural (e.g., Maple's role of a "silent moderator"). Furthermore, the actual role played by different agents in the session could not have been predicted beforehand. The process of constructing shared meanings seems to generate its own dynamics, which depends on the situation, the tasks and the available agents.

REFERENCES

Dreyfus, T. & Hadas, N. (1996). Proof as answer to the question why. *Zentralblatt für Didaktik der Mathematik 28*(1), 1–5.

Dreyfus, T. & Hillel, J. (1998). Reconstruction of meanings for function approximation, *International Journal of Computers for Mathematical Learning, 3*(2), 93–112.

Dreyfus, T., Hershkowitz, R. & Hillel, J. (2001). The emergence of algebraic structure. In H. Chick, K. Stacey, J. Vincent & J. Vincent (Eds.), *Proceedings of the 12th ICMI Study Conference: The Future of the Teaching and Learning of Algebra* (Vol. 1, pp. 213–220). Melbourne, Victoria (Australia): The University of Melbourne.

Hershkowitz, R., Schwarz, B. & Dreyfus, T. (2001). Abstraction in Context: Epistemic Actions. *Journal for Research in Mathematics Education, 32*(2), 195–222.

Hillel, J. (2000). Modes of description and the problem of representation in linear algebra. In J.-L. Dorier (Ed.) *On the Teaching of Linear Algebra* (pp. 191–207). Dordrecht (the Netherlands): Kluwer.

Janvier, C. (1978). Translation processes in mathematics education. In C. Janvier (Ed.), *Problems of representations in mathematics learning and problem solving* (pp. 89–122). Hillsdale, NJ (USA): Lawrence Erlbaum.

Pirie, S. & Kieren, T. (1994). Beyond metaphor: Formalizing in mathematical understanding within constructivist environments. *For the Learning of Mathematics, 14*(1), 39–43.

Schoenfeld, A., Smith, J. & Arcavi, A. (1993). Learning: The microgenetic analysis of one student's evolving understanding of a complex subject matter domain. In R. Glaser (Ed.), *Advances in Instructional Psychology*, Vol. 4, 55–177. Hillsdale, NJ (USA): Lawrence Erlbaum

ANNA SIERPINSKA

DISCOURSING MATHEMATICS AWAY

Studying language, communication, and meaning is like drowning in a marsh. Whatever you grab to get out of there is part of the marsh and only draws you deeper into it. It is therefore hard to explain why anyone would even want to venture into this domain. Except that, in mathematics education, one doesn't really have a choice. Any project of teaching and learning includes problems of communication. If mathematics is the object of communication, language becomes a problem. In teaching a child to eat with a spoon, walk, or ride a bicycle, language is not necessary. One usually guides the child's body and demonstrates these actions. But mathematical thinking cannot be demonstrated directly, and one cannot physically guide anybody in this activity. Communication is necessarily indirect, mediated by a combination of ordinary and highly specialised artificial languages and other sign systems. And there is no direct way of making sure the intended meaning is not lost in the mediation.

In the last twenty years or so mathematics education has struggled with (at least) three broad theoretical approaches to language: language as a code (e.g., Laborde, 1982), language as representation (e.g., Janvier, 1986; Duval, 1995), and language as discourse (e.g., Kieran, Forman & Sfard, 2001). It would be useful to start developing a unified approach to language and communication for mathematics education based on lessons learned from these experiences. This challenge would require a concerted effort of a large group of mathematics educators. This paper is a modest contribution in this direction. I tried to systematise my understanding of one of the three approaches (the discursive approach) and see how it could be linked to other approaches and results. I focused on only two texts about the discursive approach to mathematics education, namely, "FLM18.1" (Sfard in Sfard, Nesher, Streefland, Cobb & Mason, 1998) and "ESM46" (Kieran, Forman & Sfard, 2001). I shall not refer to individual authors in these texts, because my presentation and discussion of this approach are concerned with texts and not with people who have written them at some point in their lives and may have changed their views since then.

The paper is composed of two parts. In the first, I present the discursive approach as a *program* in mathematics education. The second part contains a commentary on this understanding and an exploration of the possibility of making connections with other approaches and results.

1. DISCURSIVE APPROACH AS A PROGRAM IN MATHEMATICS EDUCATION

The discursive approach to mathematics education, as presented in FLM18.1 and ESM46 (denoted by "DA" in the sequel) could be seen as a "program" in mathematics education in the sense of Sierpinska (1996). It promotes a certain *ideology* and encourages a particular *didactic action*, while developing a *theory* and conducting experimental research guided by this theory. I describe DA by identifying these three facets of the program.

1.1 The ideology of DA

1.1.1 A discursive worldview

DA views the world through the eyes of a semiotician. The basic assumption is that everything that matters is a sign, and, conversely, to make something matter, it must be turned into a sign (see, e.g., Lotman, 1990, p. 5). DA stresses the power of discourses to create phenomena and to change people's perception and experience of phenomena: "objects of thought, discourse and social manipulation are semiotic and thus culturally constituted entities" (Parmentier, 1985, p. 376; see also Bourdieu, 2001,e.g., pp. 74–75). Discourses affect how people perceive themselves; in their own eyes, the meaning of what they say and do is determined by their position (their "voice") within a society, and that is also how they interpret the behaviour of others.

1.1.2 A discursive philosophy of mathematics

Mathematics is seen as a historically developed socio-cultural practice (ESM46, pp. 66, 72); it is a "historically developed practice, dealing with certain types of objects, tools and rules" (pp. 66, 72). Mathematics was seen as a "*well-defined* type of discourse" in (FLM18.1, p. 50), but this radical view was replaced by a more moderate one in the more recent writing. DA admits the existence of not just one "well-defined" discourse of mathematics but of a large variety of mathematical discourses among mathematicians, in various professions, and in schools. It is acknowledged that school produces its own school mathematical discourses that need not be, and most of the time are not, identical with mathematical practices outside of school (ESM46, pp. 100–101).

1.1.3 A model of the learner of mathematics

DA endorses the view that "all our thinking, with mathematical thinking being no exception, is essentially discursive" (FLM18.1, p. 50). The learner of mathematics is thus an *apprentice* of mathematical discourse, represented by the teacher and textbooks. DA is aware of the difficulty this approach has in accounting for individual creativity and change of cultures (ESM46, p. 93). DA conceives of an individual as a "collection of multiple subjectivities, through the many overlapping and separate identities of gender, ethnicity, class, size, age, etc." (p. 105). The learner is therefore conceived of not as a person or a psychological subject with his or her idiosyncratic cognitive and emotional functioning, but as a member of a social group, a community with a background culture and history. It does not matter what the learner thinks to himself.[1] The learner's behaviour is interesting only insofar as it is interpreted by other members of the group as a sign; that is, as having some meaning for them.

1.1.4 A model of the teacher of mathematics

The teacher of mathematics is a *representative participant* of a mathematical culture—that is, of a discourse and praxis of mathematics—who is supposed to create conditions for the initiation of students into this culture (ESM46, p. 74). The teacher acts as a participant in his own right in the classroom discourse but also directs, "canalises", the discussion towards the relevant mathematical ideas, as well as helps in the formulation of ideas, by "revoicing" students' utterances (p. 75).

1.1.5 A model of the mathematics classroom

The class is a community; namely, a community of mathematical *discourse* (not only a community of mathematical activity, since it includes reflection and debate about mathematical activity; ESM46, p. 71). Students work on solving common problems, agreeing on approaches and techniques through social interaction, conversation, discussion, and other forms of communication. DA calls for replacing the dialogue among pupils in the classroom by "a polylogue" in a mathematical community or a "polyphonic discourse among all possible voices that helped to create the history of that community of practice" (p. 74).

1.1.6 A model of a curriculum

The program does not specify the preferred content of teaching. It only stresses that the content should be relevant from the historical and cultural point of view (it should be "historically rooted"; ESM46, p. 72). The most important objective of teaching mathematics is to initiate students into historically developed ways of doing and talking and get them involved in the mathematics "speech genre" (p. 72).

1.2 Didactic action

DA is not satisfied with the present trend of classroom conversation for the sake of conversation alone. It requires that the art of mathematical communication be an object of teaching in its own right: "communication skills cannot be taken for granted" (FLM18.1, p. 51). The implicit "meta-discursive" rules of the mathematical "genre" need to be inculcated in the learners through practice so that they become a *habitus*, a kind of second nature (ESM46, pp. 29–30, with reference to Bourdieu, 1980).

[1] The masculine forms of pronouns are used to alleviate the reading of the text.

According to (ESM46, p. 74), didactic action that could be considered as representing the discursive approach can be found in the recent work of Cobb and his collaborators (e.g., Cobb, Boufi, McClain & Whitenack, 1997; Cobb in FLM18.1), and in the Russian "dialogue of cultures" schools. I could perhaps add (with caution, since this reference does not appear in the DA texts) Boero and his collaborators' (Boero, Pedemonte, Robotti, 1997; Boero, Pedemonte, Robotti, Chiappini, 1998) experiments of introducing children into the practice of theoretical knowledge based on the Vygotskian notion of scientific concepts and the Bakhtinian notion of "voice".

1.3 Theoretical foundations

In the discursive approach, language, communication, discourse and thought do not constitute separate objects of theoretical reflection. All are included in communication: The genetic roots and purpose of language are communication; discourse is any specific instance of communicating; and thinking is a kind of communication, namely, communicating with oneself. This perspective is dictated by the assumption that the subject of acts of communication, discourse, and thinking is not a psychological individual or a person, but a social group of participants in a common culture or a collection of multiple subjectivities.

1.3.1 The meaning of "discourse"

According to Webster's dictionary, *discourse* can mean "the capacity of orderly thought; rationality", or "verbal interchange of ideas, especially conversation", as well as, "a formal orderly and extended expression of thought on a subject", and "connected speech or writing". Interestingly, "discursive" speech or writing can mean both "moving from topic to topic without order" and "proceeding coherently from topic to topic" or "marked by analytical reasoning". Thus, *discourse* in ordinary use covers all kinds of talk or writing, from casual conversation to scholarly argumentation. One thing that seems to be always required of it, however, is that it refers to something other than itself: it is talk or writing *about* something. Reciting the alphabet would not be an instance of discourse, because it does not refer to anything outside the language.

In DA, *discourse* is defined as "any specific instance of communicating, whether diachronic or synchronic, whether with others or oneself, whether predominantly verbal or with the help of any other symbolic system" (ESM46, p. 28).

1.3.2 Communication is a socio-cultural practice

DA rejects the classical sender-receiver communication model (ESM46, p. 66). It focuses not on transmission of information from one individual to another but on the participation in an activity of "sharing communalities and constructively dealing with the meanings people seem to have in common" (p. 67).

1.3.3 Language is not just code

The classical model of communication was based on the notion of language as a code; the sender would encode his thought in some symbolic form, and the receiver would decode it into his own thought. According to Lotman (1999), one of the many critics of this model,

> The term *code* refers to an idea of a structure just created, artificial and introduced based on an agreement made at some point in the process of communication. A code has no history [...] On the other hand, the term *language* unconsciously evokes in us the thought about the continuity of being. Language is a code plus its history. (pp. 31–32, my translation from Polish)

DA backs up its position by reference to Davydov's (1997/1991) writings:

> Mental functions are essentially seen as not rooted in the individual, but in the communication between individuals, in their relationships between each other and in their relationships with the objects created by people. (cited in ESM46, p. 67)

For Davydov, the subject of thinking is a "collective"; that is, a group of people participating in a common task. The collective thinking, memory, and other so-called higher mental functions are assumed not only as social in a synchronic and contextual way, but also as historical and cultural. Davydov's (1990) position is that of the dialectical materialism orthodoxy, based on the writings of Marx, Engels, and Lenin. It looks at the society from afar; individual members of the "human society" are not distinguishable the way they are in a "civil society". Sometimes DA endorses this point of view, and sometimes it is more in tune with Vygotsky's position, which is closer to the ideology of "civil society", looking at the society from the perspective of its members and speaking about the mental functioning of an individual.

Vygotsky was less concerned with applying dialectical materialism in developing his theory than Davydov was. Vygotsky still focused on the individual. He was looking at how the intermental functioning advances the intramental functioning, a perspective that led him to the notion of the zone of proximal development (Wertsch, 1991, p. 28). He saw culture and society not as "factors" that *influence* the development of the individual but as an environment with which the individual *interacts*. This environment acts on the individual as much as the individual acts on the environment (see citation from Vygotsky in ESM46, p. 67). Thus he appeared to view society as an emergent of the many interactions between participants and not, as in the dialectical materialist philosophy endorsed by Davydov (1990), a super-entity that "*confers [...] the historically developed forms of [...] activity*" on the individual (p. 232).

Vygotsky's "fundamentally a-social theory of psychological tools" (as deemed by Clot, 1999, cited in ESM46, p. 67) has been unsatisfactory for DA, which then turned its attention to Bakhtin's notion of "genre" as more "inherently social" (p.

174). The importance of being initiated into a speech genre for effective communication is highlighted in the following citation:

> If speech genres did not exist and we had not mastered them, if we had to originate them during the speech process and construct each utterance at will for the very first time, speech communication would be almost impossible" (ESM46, p. 69).

This duality of perspectives (Davydov and Bakhtin, on the one hand, and Vygotsky, on the other, i.e., the "human society" and "civil society" perspectives) leads DA to interpret the idea that "thinking is communicating" both as a methodological principle and an epistemological and ontological stance. If one takes the human society point of view, one does not need the notion of thinking at all. What is studied and what matters for the socio-cultural development of knowledge is not what this or that individual thinks to himself but what the social group decides to do and how it decides to formulate its conclusions and solutions in a process of communication. This is how the assumption "thinking is communicating" could be understood.

This methodological interpretation, however, is not quite explicit in DA, where references to Davydov and Marx coexist with speaking of thinking as communicating as an actual activity of the psychological subject. With reference to Vygotsky, DA claims that all human activity has social origins and that communicational public speech developmentally precedes inner private speech. This assumption is then claimed to imply (as in Wertsch, 1991) that thinking is a case of communication, namely communication with oneself (ESM46, p. 26). This is not meant to say that all thinking is verbal, though: "The word 'communication' is used here in a very broad sense and is not confined to interactions mediated by language" (ESM46, p. 26). How broad this sense is assumed to be is not easy to grasp. It seems that the only characteristic retained is the communicational *intent* with the aim of being *effective* (pp. 27, 28, 32):

> When one is looking at cognition as a form of communication, an individual becomes automatically a nexus in the web of social relations. *[...]* This is true whether this individual is in real-time interaction with others or acts alone. *[...]* Further, from the proposed vision of cognition it follows that thinking is subordinated to, and informed by, the demand of making communication *effective*. (p. 27; my emphasis)

1.3.4 Learning is initiation into a discourse

DA assumes that the main motive for learning (i.e., changing one's previous conceptions) is "to adjust one's discursive uses of words to those of other people" (ESM46, p. 49), especially if these other people are viewed as superior in a social or intellectual sense. Thus, "*discursive* conflicts" (p. 48) are what drives learning and not *cognitive* conflicts as assumed in the Piagetian theory of equilibration of cognitive structures.

"Learning mathematics [is] defined as an initiation [into] mathematical discourse" (ESM46, p. 28). The learner must learn two things: the means of communication and the meta-discursive rules. The latter regulate the flow of communication; they are considered akin to concepts such as Wittgenstein's *language games,* Goffman's *frames,* Bruner's *formats,* and Bourdieu's *habitus.*

"Means of communication" are described as "shaping the content of the discourse" (ESM46, p. 28). A natural question is, means of communication of... what? Normally, one would say, "means of communication of thoughts", but in this theory, thinking itself is communication, and speaking about expressing a thought whose existence would be independent of the means of communication is considered meaningless (pp. 27, 29). If so, then perhaps one shouldn't speak about the content of communication at all, but only about the discursive means of communication, the patterns of communication, and the genre of the discourse.

1.3.5 A participation model of teaching

DA promotes more an ideal of a mathematics teacher than it proposes a theory of the profession of teaching mathematics. The "participation model" of teaching presented in ESM46 (p. 117) seems the closest to a conceptualisation of teaching, but it is still put forward as an ideal model to follow (an alternative to the so-called acquisition model, and a model recommended by a reform movement). This conceptualisation of teaching also captures the teacher-in-action in the classroom but is silent about the work of the teacher when preparing for the next class and making choices regarding tasks and didactic actions based on the teacher's knowledge of mathematics, interpretation of the curriculum, and evaluation of the happenings and progress in the previous classes. It would be a real methodological challenge to develop a "discursive approach" to investigating teacher's thinking and action between classes. For the time being, the teacher's work in all its phases has been an object of investigations from much broader perspectives (e.g., Chevallard, 1999; Coulange, 2001; Salin, 2002).

Concerning ways in which a teacher can initiate students into the practice of mathematics, DA asserts, based on empirical evidence, that indirect methods ("co-constructive creation of mathematical models", ESM46, p. 81) are more effective that teaching readymade models.

2. COMMENTARY

DA appears to be part of an epidemic of "discursive approaches" in the social sciences. The label can be found in psychology and psychiatry (e.g., Edwards & Potter, 1992; Harré & Gillet, 1995; Gillet, 1999), sociology (Bingham, 1994), political science (Dryzek, 1990). The meanings of this expression, however, vary from domain to domain, and they have changed over the years. Twenty years ago, the "discursive approach" in education may have meant using essay writing, discussion, and debate forums as forms of communication in a whole range of subjects at school (Chilver, 1982). In political science, "discursive democracy" has been defined as an alternative to "instrumental democracy" (Dryzek, 1990). In sociology, "discursive approach" may refer to critical approaches that are concerned with the problems of shaping social, cultural, and political realities by certain discourses; not only analysing the discourses but also proposing political action to change the realities for culturally discriminated groups.

2.1 Cultural conformism

DA defines itself in opposition to "cognitivist" or "individualistic" approaches in psychology and in opposition to the "acquisition" metaphor in teaching methodology. But it is not a "critical theory", in the sense that it does not challenge using mathematics as a selection tool in the education system (Bourdieu, 1994, p. 49)[2], and it does not question the existing school mathematical discourses to the point of wanting to transform them in directions that would allow the "economically disadvantaged groups" to perform as well as others. The statement that these groups do not perform as well as others seems to be taken as a fact (ESM46, p. 94). It would be interesting to investigate what percentage of mathematicians who made significant contributions to their field came from economically disadvantaged groups and to what they owed their success. Was it indeed their exceptional facility to become *initiated into the existing mathematical discourses,* or was it rather their ability to grasp the main mathematical ideas and build new ones from them, constructing novel mathematical discourses? Srinivasa Ramanujan and Stefan Banach, to name only two famous mathematicians raised in poverty, would fall into the latter category.

2.2 "Discursive conflict" and "socio-cognitive conflict": The social motives of learning

DA proposes to replace "cognitive conflict" by "discursive conflict" as a motive for learning. The notion of discursive conflict appears close to the notion of "socio-cognitive conflict" introduced over twenty years ago in social psychology (Doise & Mugny, 1981; Perret-Clermont, 1979), also in reaction to the limitations of the notion of cognitive conflict. The socio-cognitive conflict can be seen as a discursive conflict; it was understood as the co-occurrence of contradictory statements in a situation of social interaction (Blaye, 1989, p. 186). The notion of socio-cognitive conflict had been thoroughly researched and studied from several theoretical points of view (the cognitive point of view of Piaget, the socio-cultural perspective of Vygotsky, and theories of social learning, e.g., Bandura, 1980). Results of this research could caution DA researchers against trusting some of their conjectures. The conditions of progress through socio-cognitive conflicts are not obvious, and they are complex—not reducible to the "model effect" or wanting to adjust one's use of words to that of someone considered more knowledgeable (see, e.g., Mugny,

[2] "C'est souvent avec une grande brutalité psychologique que l'institution scolaire impose ses jugements totaux et ses verdicts sans appel qui rangent tous les élèves dans une hiérarchie unique des formes d'excellence—dominée aujourd'hui par une discipline, les mathématiques". (Bourdieu, 1994, p. 49)

Doise & Perret-Clermont, 1975-1976). It was found that interactions, which could be regulated otherwise than by a *co-elaboration* of a solution (e.g., by the social or other authority of one of the partners), were not efficient in generating progress. Moreover, it was not possible to confirm experimentally a greater efficiency of inter-individual conflicts over intra-individual conflicts (children were doing better in a social conflict situation only if they did not have the possibility of verifying their hypotheses experimentally; Blaye, 1989, p. 190).

2.3 "The regulating effects of discourses" and "institutional contracts": The impact of the didactic contract on what is learned in school

The hypotheses proposed in social psychology have inspired mathematics educators to study the social dimensions of learning mathematics in more natural situations of ordinary classrooms. Extensive research in this direction was conducted using a variety of theoretical frameworks (interactionist, socio-cultural, theory of didactic situations, anthropological approach; see, e.g., Venturini, Amade-Escot & Terrisse, 2002). Whereas interactionist approaches stressed the verbal interaction patterns among students and the teacher in a classroom, researchers starting from what is now called an "anthropological approach" in France looked more at the implicit institutional contracts regulating the mutual positioning and behavior of the participants. Already in 1986, Balacheff was explaining the difficulty of obtaining the engagement of students in mathematical proving by the social character of the classroom situation. In this situation, most of the time, the student acts as a practical person, not as a theoretician; he aims at producing a solution (a text) that would be acceptable for the teacher, not at producing knowledge (see also accompanying comments in Sierpinska, 1994, pp. 18–19).

DA assumes that the social situation of the classroom, the relations of the assumed epistemological authority of the teacher, and the teacher's responsibility for the students' learning are factors that allow students to learn "what counts in the community's speech genre" (ESM46, p. 75). They may indeed learn all that, but they may also learn that it is the *teacher's* and not *their* responsibility that they learn and that what they claim is mathematically valid. The teacher's feeling of epistemological *responsibility* may result in the students' feeling of *dependence*, and that may happen even under the conditions of an apparently adidactic situation, with students supposed to be working in small groups on a relatively open problem and even in computer labs (Arsac, Balacheff & Mante, 1992; Bellemain & Capponi, 1992; Laborde, 1992, pp. 1–4). Students' dependence on the teacher for checking the validity of their solutions is part of the didactic contract, a specificity of the school institution. In these conditions, it is not realistic to expect that students will be "initiated" into the discourse of working mathematicians; they can only be initiated into the discourse of *school mathematics*, where proofs are texts written in a distinct genre, having little to do with the truth of statements the students are supposed to prove and even less with communication of a result of an investigation.

2.4 *Taking into account the social, cultural and historical sources of knowledge in analyses of communication episodes*

The above argument is not to say that school mathematics is bad mathematics and that it should be replaced by "genuine" mathematics. Mathematics at school will always be school mathematics as long as schools are schools and not research institutes or other workplaces. On the contrary, school mathematics must be accepted and taken seriously. For research in mathematics education, this implies that, whenever an "episode" of group or classroom communication is analysed, the history and culture of the classroom in which it took place, as well as the experimental contract between the observed students and the researchers, must be included not only as "background information" for the reader but as important data to be analysed. Part of this history is the design of the tasks given to the students. These tasks were designed relative to some goals and socio-cultural constraints. It is important to factor in these data in an interpretation of the episode. It would be natural for DA, with its focus on the socio-cultural and historical roots of knowledge, to apply this principle to its own research. It was rather unexpected, therefore, that the presentations of episodes provided in ESM46 were extremely detailed at the level of tape-recorded and transcribed verbal exchanges, but very cursory at the level of the socio-cultural context of these exchanges, the histories of the participants, and information about the design of the tasks. I could only agree with Hoyles' (2001) commentary on the ESM46 texts: "What I missed was any discussion of the *design* of the activities and the *design* or *choice* of the tools or sign systems that were introduced to foster *mathematics* learning" (p. 284).

2.5 *Collectivisation of thought*

The slogan in the 1960s was: individualisation of teaching. Now the slogan is collectivisation of teaching. And even, collectivisation of thought. Nobody can expect this author, who had the opportunity to live through the realisation of this ideology in socialist Poland, to take the idea of collective thought seriously. The official political discourse was full of statements such as "the collective of [name of factory] expressed their full support for the leading role of the party". This usually meant that the attempt of the workers to go on strike was unsuccessful.

According to Wertsch (1991), Vygotsky's and other Soviet psychologists' theories were an attempt to apply Marx's thesis that "human's psychological nature represents the aggregate of internalised social relations that have become functions for the individual and form the individual's structure" (p. 26). Translating this ideological assumption of social and cultural roots of human thinking, voluntary attention, and other higher mental functions into a psychological theory and supporting it with evidence from empirical studies afforded political activists a scientific justification for decisions that were not so scientific. Reality had to be made to fit the theory, and if it didn't, so much the worse for the reality. Children had to become participants of the socio-cultural community and initiated into the model discourse at an early age; they were sent to a nursery and then to a kindergarten, where they learned the politically correct discourses of the time. At work, adults were part of a collective, which then collectively made decisions, formulated resolutions, and generally never expressed any individual original thought that would go "against the will of the socialist society". Anybody expressing such individual original thought was immediately labelled an "enemy of the people" and weeded out from the "healthy body of the socialist society".

2.6 Smooth but meaningless communication or difficult but meaningful communication? Language – genre = code

Bakhtin said that without a mastery of speech genres, "speech communication would be almost impossible" (cited in ESM46, p. 69). *Almost* impossible but not completely impossible: This claim is supported by the possibility of communication between people coming from different cultural backgrounds and speaking to each other in a language that is foreign for at least one of them. If someone has learned a foreign language late in life, it is a *code* for that person—a language without history, without a genre. A dialogue with this person can be very difficult but not impossible, and all the more significant. If the interlocutors use the same familiar genre, communication is smooth but trivial: They have nothing to say to each other (Lotman, 1999, p. 32). High level of information obtains in cases of difficult translation between the utterances of the interlocutors (p. 33)[3]. It may well be that progress in science owes more to difficulties in communication than to efficient communication. One may conjecture that research done by a multilingual team is more likely to be innovative than if the team shares the same "genre" (and jargon). Can the same be conjectured about multilingual mathematics classes (see Adler, 2001; Sierpinska, 2002)?

2.7 Problems with defining communication by intent and effectiveness

In view of DA's rejection of the classical model of communication, its focus on intent and effectiveness is rather unexpected. The classical model also defined communication by intent and effectiveness. This invited many theoretical difficulties (see, e.g., Bruner, 1974, p. 262; Lotman, 1999, p. 31). How could one establish whether the interpretation of a message was indeed the one consciously intended by the author of the message? If that is impossible, then how can we tell whether communication was effective or not? How can we even tell whether an utterance was an instance of communication (did the speaker intend to communicate)?

[3] A similar idea was expressed by Lotman (1990, pp. 80–81): "We should not [...] forget that not only understanding, but also misunderstanding is a necessary and useful condition in communication. A text that is absolutely comprehensible is at the same time a text that is absolutely useless. An absolutely understandable and understanding partner would be convenient but unnecessary, since he or she would be a mechanical copy of my 'I' and our converse would provide us with no increase in information."

> Intent in communication is difficult to deal with for a variety of reasons, not the least demanding of which is the morass into which it leads when one tries to establish whether something was *really*, or *consciously* intended. *[...]* To obviate such difficulties, it has become customary to speak of the *functions* that communication or language serve and to determine *how* they do so. This has the virtue, at least, of postponing ultimate questions about 'reality' and 'consciousness' in the hope that they may become more manageable. (Bruner, 1974, p. 262)

A solution was proposed by Roman Jakobson, who modified the classical sender-receiver model of communication to include the "context" while shifting the focus from the "vertices" of the structure (i.e., sender, receiver, message, context) to the relations amongst them. These relations were conceptualised as functions of language in communication: emotive/expressive, poetic, conative, phatic, metalinguistic, and referential (Jakobson, 1960; see also a discussion and an application of this model to language acquisition in infants in Bruner, 1974). All these functions of language were assumed to be there in any act of communication; different acts of communication differed only by a hierarchical order of these functions. Jakobson's categories can be quite useful in modelling the use of language in the teaching and learning of mathematics, as I hope to demonstrate in a future publication (Richard & Sierpinska, forthcoming).

2.8 Relations between discourse, communication and thought

Based on Vygotsky's assumption of social origins of higher mental functions, DA conceptualises thinking as communicating—namely, as communicating with oneself—and claims that all our thinking is discursive. This conceptualisation does mathematical thinking no justice, and it doesn't seem useful for the purposes of mathematics education. Arguments against it, however, cannot be based on the abundant empirical evidence that thinking, especially mathematical thinking, is not all verbal, because DA claims that communication (and discourse) is not necessarily verbal and may be based on other signs, such as images. The following arguments may apply, however:

1. Communication is a voluntary act; not all thinking is voluntary.
2. Mathematical thinking requires well-developed spatial visualisation (not only in geometry but also in dealing with numbers and algebra); spatial visualisation is seeing things and moving spatial configurations around in one's mind; "seeing" is not intentional pointing things to oneself and therefore cannot be regarded as an act of communication.
3. Although "higher mental functions" are often analytic—that is, mediated by conventional sign systems and therefore with social origins—they are based on brain activity that has biological origins preceding the use of conventional sign systems both in individual development and in the evolution of the species.
4. Not all instances of communication are discursive.
5. Discursive thinking is not necessarily communicational.

Let me elaborate on these points.

2.8.1 Involuntary thinking

We seem to think on several distinct planes simultaneously, which may correspond to activity in different parts of the brain. There is certainly a plane where thoughts just *happen to us* whether we want it or not: It is the plane of involuntary (but not necessarily unconscious) thinking. This thinking may contain words or images, but neither the words nor the images are intended for communication. This involuntary thinking can be responsible for the moments of "illumination", where the solution of a problem is suddenly revealed. Accounts of this phenomenon in mathematicians can be found in Hadamard (1945).

2.8.2 Spatial visualisation in mathematics

Dynamic spatial visualisation plays an important role not only in geometric thinking but also in algebra. Let me explain through the following example.

Suppose I am asked to prove that if A is any real $n \times m$ matrix, then the vector space R^m can be decomposed into a direct sum of the null space of A and the range of the transpose of A:

$$R^m = N(A) \oplus R(A^T)$$

Suppose I see the situation represented by this equation as being about transformations: A transforms R^m into R^n, and the transpose of A transforms R^n into R^m. The null space of A is the part of R^m that is transformed into the zero vector of R^n (I see it shrinking to a point). The range of A^T is the part of R^m composed of all vectors obtainable as $A^T y$ with y in R^n. As I am saying all this, my mind works as if looking from left to right and back. I keep some imaginary places for R^m (on the left) and R^n (on the right), and I move from one to the other. I see the task in front of me as understanding why the space on the left can be seen as generated from these two parts. My prototypical image of a direct sum being something like a space spanned on two perpendicular axes, I immediately see these two parts as being in a horizontal-vertical position. Hence the idea of proving that the orthogonal complement of $N(A)$ is $R(A^T)$ (since $V = W \oplus W^{\perp}$ in general, this does the proof). Thinking this way, I am moving from left to right, going from R^m to R^n by means of multiplication by A. But this turns out to be difficult. When I write the orthogonal complement of $N(A)$ (written $N(A)^{\perp}$), by definition, $N(A)^{\perp} = \{y \in R^m$: if $Ax = 0$ then $<y, x> = 0\}$, I don't see at all why that should be equal to the range of A^T, that is, the space generated by the columns of A^T or the rows of A. When thinking about that, my mind constantly moves between viewing the matrix A vertically, as a set of columns, to viewing it horizontally, as a set of rows, and then transposing the configuration to see which is which in A^T. All these gymnastics lead nowhere until I suddenly change my point of view and start looking from right to left, from R^n to R^m using A^T, and from below to the top, from $R(A^T)$ to $N(A)$. Instead of trying to show that $N(A)^{\perp} = R(A^{T})$, why not try to show that $R(A^{T})^{\perp} = N(A)$? Realising that two transpositions return the matrix to its original position (again some mental gymnastics in moving from vertical to horizontal positions), I see that it is enough to prove that $R(A)^{\perp} = N(A^{T})$ and then apply it to the transpose of A. Equality may be a reflexive relation, but as I am thinking about this proof, I go from the left to the right of the equalities I write, and I am thinking of $R(A)^{\perp} = N(A^{T})$ as saying that the orthogonal complement of $R(A)$ is the same as the null space of the transpose of A, and not that the null space of the transpose of A is equal to the orthogonal complement of the range of A. The former is the first thing one notices; the latter is not so obvious.

I stop my story here. I hope it conveys the importance of the "muscular" mental effort involved in a very simple case of mathematical reasoning about matrices.

Spatial visualisation expresses itself also in the sense of rhyme and rhythm, responsible for mathematicians' proverbial "perfect pitch" for patterns. There are innumerable instances of this phenomenon at any level of mathematical sophistication. For example, our preference for writing polynomials in one variable so that the powers of the variables are arranged in increasing or decreasing order rather than in random order can be seen as a source of some mathematical notions; for example, the isomorphism between the vector space of polynomials of degree ≤ n and R^{n+1}. Also, indexing the coefficients of a general quadratic form so that they match the variables brings forth the idea of the matrix representation of a quadratic form (Figure 1).

$$a_0 + a_1x + \ldots + a_nx^n \leftrightarrow (a_0, a_1, \ldots, a_n)$$

$$a_{11}x_1^2+2a_{12}x_1x_2+2a_{13}x_1x_3+a_{22}x_2^2+2a_{23}x_2x_3+a_{33}x_2^2 = \begin{bmatrix} x_1 & x_2 & x_3 \end{bmatrix} \begin{bmatrix} a_{11} & a_{12} & a_{13} \\ a_{12} & a_{22} & a_{23} \\ a_{13} & a_{23} & a_{33} \end{bmatrix} \begin{bmatrix} x_1 \\ x_2 \\ x_3 \end{bmatrix}$$

Figure 1: Notational rhythms and rhymes

$$x_1 = 5 + 6x_2,\ x_4 = 4x_3 - x_2,$$

$$x_2 = 4x_4 + 5x_3\ ,\ x_3 - x_1 = 5x_4 - 6x_2 + 3$$

$$\begin{bmatrix} 1 & -6 & 0 & 0 \\ 0 & 1 & -4 & 1 \\ -1 & 6 & 1 & 5 \\ 0 & -1 & 5 & 4 \end{bmatrix} \begin{bmatrix} x_1 \\ x_2 \\ x_3 \\ x_4 \end{bmatrix} = \begin{bmatrix} 5 \\ 0 \\ 3 \\ 0 \end{bmatrix}$$

Figure 2: A rhythmic representation of a linear system leads to its matrix representation

Given a concrete set of linear equations, one can always solve it by isolating variables from the simpler equations and substituting them into the more complicated ones. But this way of working is messy and specific to each concrete case (Figure 2). It is by trying to give some rhythm to the text of the solution that we arrive at the matrix representation and a general method of solution.

It is this sense of rhyme and rhythm that could have led Cayley and Sylvester to the development of matrix theory and that could have supported their study of algebraic invariants and linear transformations, which, at that time, were linear substitutions of variables (see, e.g., Bell, 1956).

> Sylvester's sense of the kinship of mathematics to the finer arts found frequent expression in his writings. Thus, in a paper on Newton's rule for the discovery of imaginary roots of algebraic equations, he asks in a footnote 'May not Music be described as the Mathematic of sense, Mathematic as the Music of reason? Thus the musician feels Mathematics, the Mathematician thinks Music—Music the dream, Mathematic the working life-each to receive its consummation from the other when the human intelligence, elevated to its perfect type, shall shine forth glorified in some future Mozart-Dirichlet or Beethoven-Gauss—a union already not indistinctly foreshadowed in the genius and labours of a Helmholtz!' (Bell, 1956, p. 364)

2.8.3 Biological roots of mathematical thinking

According to Vygotsky, thought and language develop separately till the age of two, when their paths start intertwining. At a very early age, 0-2 months, infants start noticing things, recognising faces, and they start babbling. These are the beginnings of thinking and speech, but neither has some hidden communicative intent. Babbling fulfils the same function for the tongue and vocal cords as moving arms and legs do for the development of muscular strength. Infants babble and move without any intention of communication; they do it just because they are "alive and kicking". However, the caretakers react to the babble as an invitation to engage in communication, and they "talk back" to the infant. It seems that this is how children learn that the noises they are making can be used for making and maintaining contact with others. Although certainly the use of language to maintain social contact with others is quite important in children's and adult's lives, the original biological roots of language reappear in its use as an activity in itself. Reflection about language and other means of communication is specific to humans. Thus, although at some early (but not too early) stages of language development, "the purpose of language is communication" (Bruner, 1974, p. 276), later language may be used as material in the hands of an artist or a scientist.

Objects of scientific knowledge owe their socio-cultural existence to linguistic forms of communication, but these linguistic forms do not completely determine their nature, because the non-discursive modes of thinking may twist and turn them around in discursively unpredictable ways. The role of non-discursive modes of thinking is difficult to identify in research because of their implicit character, and it is not easy to prove their evolutionary precedence over discursive thought. Such attempts have been made, however. For example, Corballis (1997) studied brain functioning during mental rotation of shapes, a mental activity considered to be "a paradigmatic example" of "a higher-order process that is non-symbolic and analog as opposed to propositional". Corballis observed that, although this activity engages one hemisphere (the right one) slightly more than the other, this bias is nothing compared to the concentration of the brain activity in the left hemisphere during language processing. This observation led Corballis to formulate the following hypothesis:

> The characteristically symbolic mode of the left hemisphere evolved relatively late and achieved the quality of recursive generativity only in the late stages of hominid evolution. This forced an increasingly right-hemispheric bias into analog processes like mental rotation. Such processes nevertheless remain important and integral even to those processes we think of as highly symbolic, such as language and mathematics. (p. 100)

Evidence for biological roots of mathematical thinking that stresses the importance of non-discursive modes of thinking is also provided by the phenomenon of Williams syndrome. Williams syndrome individuals are very sociable, communicating with great ease and producing grammatically sophisticated discourse in natural language. Their visual processing is severely impaired, however, and their mathematical skills remain at the level of a 7-year-old child (Rossen, Klima, Bellugi, Bihrle & Jones, 1996):

> A WMS adolescent is characteristically unable to perceive gross distinctions in orientation or to draw or copy simple stick figures. *[...]* There is selective attention to details of a configuration at the expense of the whole. (p. 375)

Given a shape made of identical small shapes—for example, a big letter D made of little Y's—Williams syndrome adolescents would perhaps notice the Y's but not that they are arranged into a big letter D. Asked to reproduce the figure, they would draw some Y's, perhaps in two rows or vertically. Although WMS adolescents use lexically rich language and correct grammar, they do not score as well as normal subjects on tasks requiring definitions of words. This finding supports the hypothesis that the biologically more primitive spatial sense is necessary even in such paradigmatically discursive tasks as defining.

2.8.4 Not all instances of communication are discursive

Following Benveniste (1966, 1974), Duval (1995) insisted on the condition that the use of language in discourse refers to something else than language itself (a condition implicit in the ordinary understanding of the word). When humans use some semiotic system of representation (a natural or a formal language, a picture, a diagram, a geometric figure, a graph, a musical sound; see Duval, 1995, p. 27) to say something "about the world" (i.e., not just to produce the sounds of a language, as in reciting an alphabet, or to play the scales or to display the colours in a palette), in a way that is shared by those who use the system to communicate, then we could say that they produce a discourse.

In this sense, when people use language in its *phatic* function—that is, for the sake maintaining contact with an interlocutor (Jakobson, 1960)—they are not producing discourse. Expressions such as "right?" at the end of an addresser's statement and the "mmm" noises of the addressee, in response to them are not saying anything "about the world". In some instances of communication it doesn't really matter what one says; it only matters that one is there.

Language in phatic function is not something specific to human language, but it is specific to communication. In fact, one could define communication as *the property of any system, made of individual elements endowed with the ability to act independently, that allows the coordination of these individual actions so that the system remains a system and does not disintegrate*. One could speak of communication among the cells of an organism, as well as of communication in social groups of animals or humans. Unlike the DA definition (see section "Communication is a socio-cultural practice"), the proposed definition focuses on the coordination of *differences* between individual participants rather than on "sharing *communalities*", which, as mentioned above, could lead to trivial exchanges.

Although DA focuses on communication in its theory and its ideology, it does not seem to attach much importance to the phatic function of language in analysing instances of communication in the practice of research. In social situations that force people to maintain communication (which is the case if the conversation is being audio- or video-recorded for the purposes of research), one or both interlocutors may use language mainly in its phatic function. What they say may have little to do with what they think, and the pressure to speak may even prevent them from thinking the way they would if they didn't have to maintain a conversation. Some students simply cannot communicate and think at the same time because their thinking may be visual or tactile or otherwise non-symbolic. They are then declared a failure in mathematics as well as in communication. Students who are sufficiently self-confident to not care about the censure may manage to speak out loud while in fact, thinking to themselves. They may produce expected solutions but not a clear and coherent discourse. They may be using wrong terms for what they think about, since they are not necessarily thinking in words. These students are then blamed for not having the necessary communicational skills, and it is recommended that they be taught the relevant meta-discursive rules.

Other examples of communication without discourse could perhaps include children's play with language, as in the inscription produced by a 6-year-old shown in Figure 3. This child was trying to imitate inscriptions including number sentences that she saw in the colourful booklets addressed to children her age and in her computer games. She was thus communicating that she "could write". But the ideas of addition and equality used in playing games and sharing food and toys were at that time separate in her mind from the symbols "+" and "=". It is perhaps the same intention to communicate "I can write mathematics" that underlies the behaviour of some undergraduate students who produce meaningless strings of set-theoretic symbols by way of "proofs". This phenomenon has been analysed as the obstacle of formalism in linear algebra students (Dorier, Robert, Robinet & Rogalski, 2000; Sierpinska, 2000).

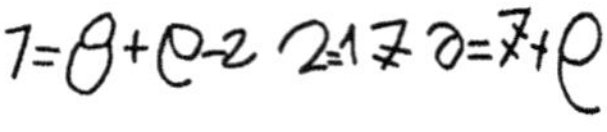

Figure 3: A "mathematical sentence" produced by a 6-year-old child

2.8.5 Non-communicational functions of discourse

Children as young as 2 years are found to engage in linguistic activities outside of a directly communicative context (Clark, 1978, p. 32). They spontaneously correct and comment upon "their own [and others'] pronunciations, word forms, word order, and even choice of language in case of bilinguals"; they make "judgments of linguistic structure and function, deciding what utterances mean, whether they are appropriate or polite, whether they are grammatical" (p. 32). Children play with language; they "play with different linguistic units, segmenting words into syllables and sounds, making up etymologies, rhyming and punning" (p. 32). For Clark, the growing sophistication of the meta-linguistic use of language is closely linked with the growth of meta-cognitive skills, such as monitoring one's ongoing utterances, checking the result of an utterance, testing for reality, deliberately trying to learn, predicting the consequences of using inflections, words, phrases or sentences, reflecting on the product of an utterance (p. 34). The last skill includes such uses of language as "providing definitions" and "explaining why certain sentences are possible and how they should be interpreted". These activities are basic in the construction of any theory, including a mathematical theory.

According to Duval (1995), communication is one of the three functions of use of language that are not specific to language but are common to all semiotic systems of representation (called "meta-discursive functions"). The other two are *processing* and *objectivation*. Formalised processing is specific to mathematics, logic, and computer science. Objectivation is the use of language in the aim of obtaining some control over one's activity and over one's experience, whether physical or mental. Objectivation organises and reorganises one's activity and experience and makes it the object of conscious evaluation and decision. It is not a mere explicitation or expression of a thought:

> The work of 'writing up', the literary creation and finding words in the frame of an analysis come under this function of objectivation first of all. But objectivation is not specifically tied with language; it can also be realized with figural semiotic systems, such as drawing, for example. [...] *The function of objectivation is irreducible to a social function of communication.* Trying to understand a discourse produced for objectivation purposes as if it was a discourse produced for communication purposes not only creates a misapprehension about what is being said but also breaks the communication

> with the author of the discourse. (Duval, 1995, p. 90, with reference to Lacan, 1966; my translation from French, my emphasis, A. S.)

DA does not distinguish among objectivation, processing, and communication. This leads to occasional misinterpretation of discourses of students who are using language for objectivation or even processing purposes as symptomatic of their lack of communicational skills.

Apart from the above meta-discursive functions of language, Duval (1995) distinguished functions that are specific to the use of language ("discursive functions"). These functions are called *referential* (naming objects), *apophantic* (making statements about the named objects), *discourse expansion* (linking different statements into a coherent whole), and *reflexive* (marking the value, the mode or the status of the expressions used). Each function can be fulfilled by means of different "discursive operations" (Grize, 1982). Discursive operations are used by the user of language to schematise or organise the discourse: Describing, explaining, and reasoning by rhetorical argumentation or by logical deduction are examples of operations used in the discourse expansion function. The idea of discursive operations may partly overlap with DA's "meta-discursive rules".

The distribution of the meta-discursive and discursive uses of language as well as of the discursive operations may be characteristic of the domain of reference of the discourse. Mathematics, for example, favours the meta-discursive function of processing, the referential discursive function, and the recurrent use of the operation of description in apophantic uses of language (Duval, 1995, p. 95).

Let me illustrate this on an example of a typical sentence in a linear algebra text: *Let T be a linear operator on a vector space V of finite dimension over the field K.* This sentence draws the reader's attention to an object and gives it a name, *T,* for further reference. The operation of description is used 6 times. *T* is a name of an operator *described* as linear, and *described* as defined on a space, *described* as vector space, further *described* as having its scalars in the field named *K* and further *described* as being of dimension *described* as finite. We could represent the recurrent use of the operation of description by means of the following diagram:

[*T* is [linear [operator]] on a [[finite [dimensional]] [[vector [space]] over a field *K*]]]

The high number of descriptions per sentence and their use in a recurrent manner are quite specific to mathematics. (Other kinds of measures that distinguish mathematical discourse from literary discourses are presented in Duval, 1995, pp. 108–110). In oral communication with students and colleagues, we often simplify the discourse and omit the descriptors. We just say, "Take an operator *T*", and the rest is understood from the context. But even if we use precise language in our lectures, students ignore the descriptors in an attempt to reduce the complexity of the discourse. For example, a statement such as,

If $U_1 = Span\ \{(1,0,1), (0,1,0)\}$, $U_2 = Span\ \{(1,0,-1), (0,1,-1)\}$,

$$\text{then } U_1 + U_2 = \text{rowspace}(\begin{bmatrix} 1 & 0 & 1 \\ 0 & 1 & 0 \\ 1 & 0 & -1 \\ 0 & 1 & -1 \end{bmatrix}) = \text{rowspace}(\begin{bmatrix} 1 & 0 & 0 \\ 0 & 1 & 0 \\ 0 & 0 & 1 \\ 0 & 0 & 0 \end{bmatrix}) = R^3.$$

often becomes, in students' rendition, something like,

$$\text{basis } U_1 = (1,0,1), (0,1,0), U_2 = (1,0,-1), (0,1,-1),$$

$$U_1 + U_2 = \begin{bmatrix} 1 & 0 & 1 \\ 0 & 1 & 0 \\ 1 & 0 & -1 \\ 0 & 1 & -1 \end{bmatrix} \sim (\text{rref}) \sim \begin{bmatrix} 1 & 0 & 0 \\ 0 & 1 & 0 \\ 0 & 0 & 1 \\ 0 & 0 & 0 \end{bmatrix} = R^3.$$

For some students, this is just shorthand, a result of a metonymy, and their concepts are correct even if their writing is rather sloppy. For other students, it could be a literal substitution of concepts: The subspaces have only two vectors each, and their sum is a matrix made of these vectors; R^3 obtains whenever an identity 3×3 submatrix is reached in row reduction. Indeed, for some students the only object of linear algebra is the matrix, and the only operation is row reduction. Instead of lamenting the students' lack of conceptual understanding, however, we could look at their activity in a positive way. The students' writing could be understood as a *non-discursive representation*—a schema or a script—of a *synoptic apprehension* of the expanded mathematical discourse, and only a first step towards a detailed understanding of the discourse (Duval, 1995, p. 354). This is a healthy and effective approach to the study of mathematics.

2.9 The importance of discursive skills for high achievement in mathematics

In our recent research on theoretical thinking in high-achieving undergraduate students (Sierpinska, Nnadozie & Oktaç, 2002), we were rather far from equating thinking with communication and mathematics with a kind of discourse. Mathematical thinking was assumed to be based on an interaction of practical and theoretical thinking. Practical thinking included visual imagery and technical skills and was considered to be the source of wonder and curiosity leading to bold conjectures, which then provided food for theoretical thought. Theoretical thinking could qualify as discursive thinking. We assumed that theoretical and practical thinking differed in their aims, objects, main concerns and results, as follows:

Aims: Theoretical thinking is thinking for the sake of thinking; practical thinking is thinking for the sake of getting things done or making things happen.
Objects: Practical thinking is thinking about particular "objects" (things, matters, events, people, phenomena). The objects of theoretical thinking are systems of concepts.

Main concerns:

a) *Meaning vs. significance*: Theoretical thinking is concerned with *meanings of concepts*, whereas practical thinking is concerned with the *significance of actions*.

b) *Conceptual vs. factual connections*: Theoretical thinking asks questions about the possible consequences of assumed meanings for the meanings of other, related concepts. The theoretical thinker does not take the meaning of concepts as personal property. Practical thinking is concerned with contingency in time and space, analogy between observed circumstances across time, particular examples, and personal experience.

c) *Epistemological vs. factual or social validity*: Practical thinking is concerned with *factual validity*. The proof of a plan of action is in the results of the action and in the agreement of the assumptions of the plan with experience, not in the internal coherence among the assumptions, the steps, and the expected outcomes. But conceptual coherence and internal consistency of systems of symbolic representations (*epistemological validity*) would be exactly the concern of theoretical thinking. Aware of its distance to experience, theoretical thinking makes no claims to stating "the truth" about experience. Theoretical thinking produces "propositions" that are conditional or *hypothetical statements*. Moreover, theoretical thinking is concerned not only with what might appear plausible or realistic, but also with what is *hypothetically possible*: Thus it tends to analyse all logically possible cases or consequences of an assumption even if they are practically unlikely.

d) *Methodological vs. technical concerns*: Practical thinking is concerned with the availability of alternative courses of action if a chosen one does not work. In a way, practical thinking always operates on a single level of its relation with its aims and objects: the level of action whose purposes are external to thinking itself. Theoretical thinking, on the other hand, operates on two levels. It reasons about concepts and it reasons about that reasoning. Theoretical thinking aims at an explicit formulation of its "methodology".

e) *Systemic vs. ad hoc approach to symbolic representations*: Theoretical thinking is concerned with symbolic notations and forms of graphical representation, and with the rules and principles of reasoning and validation that it uses. It wants notations that could be applied for expressing ideas and relations in many areas rather than only some ad hoc symbols, different in solving each particular problem.

Results: Results of practical thinking are changes in the objects of this thinking (construction of a new thing, change of a course of events, change in the behaviour of people, etc.). Results of theoretical thinking are theories and specialised notations.

We conducted structured interviews with 14 students who achieved high grades in a first undergraduate linear algebra course (see Sierpinska et al., 2002). Twelve of them completed the second algebra course in the following semester, and 6 achieved high grades in this course also. We operationalised our model of theoretical thinking in terms of concrete behaviours in responding to the interview questions and introduced three statistical indices to capture the students' tendency to theoretical thinking. Let me mention here only some findings related to linguistic and meta-linguistic sensitivity. Sensitivity to formal notation, to specialised mathematical terminology, and to the structure and logic of mathematical language was estimated at 0,71 in the subgroup of the six high achievers in both courses, compared to 0,62 for the whole group (these numbers represented the average probability that a randomly chosen student from that group would display such sensitivities). The chances that a student from the subgroup of high achievers in both courses would engage in mathematical investigations beyond the questions asked in a given problem was estimated at 0,67 (0,46 for the whole group), and the chances of spontaneously engaging in proving or hypothetical thinking were 0,58 (0,43 and 0,54 for the whole group, respectively). One of the interviewed students (coded O2) failed the second course; his sensitivity to mathematical notation and terminology was very low. These results may suggest that *being fairly comfortable with the conventional mathematical discourse is more important for high achievement in mathematics than investigative disposition or concern for the validity of one's statements*. This result is consistent with the previously cited claims that school mathematics is a special type of discourse that is quite different from mathematics as practiced by working mathematicians.

Incidentally, O2 displayed an exceptional investigative disposition in our interview. Here is an example of his behaviour in one of the questions of the interview. In this question, students were shown two lines in log-log base 2 scales, that is, the units on both axes represented 2^0, 2^1, 2^2, etc. (see Figure 4). Both lines looked straight. The upper one crossed the vertical axis at 2, and its visual slope was 1. The bottom line passed through the origin, and its visual slope was 1/2. Thus the upper line could represent a linear function (namely, $y = 2x$); the bottom line could represent $y =$ sqrt(x). Students were asked if they thought that these lines represented linear functions. The student O2 verified the linearity of the functions by calculating difference quotients for some pairs of points of the graphs. He then wondered, in relation to the bottom graph, "Looks like a linear function, but it's not". He continued thinking about the representation, trying to visualise it, and making gestures with his hands to show shapes of graphs. He proposed an interesting hypothesis. He conjectured something to the effect that lines above and below the upper line should represent "parabolas", and the only line that could represent a linear function would be the one in the position of the red line. He was making gestures to show that the lines above the red line would "curve" to the right and upwards and those below would "curve" to the right and horizontally. Indeed, the only straight lines in the log-log scales that represent linear functions are those with slope equal to 1. Let $z = at + b$ be the equation of a straight line in log-log scales with base 2. This means that $z = \log(y)$, $t = \log(x)$ and $b = \log(2^c)$, where log stands for the base 2 log. Thus $y = 2^c x^a$. This function is linear only if $a = 1$. If $a > 1$ the graph indeed "curves upwards", and if $0 < a < 1$, it "curves horizontally".

Unfortunately, the investigative disposition did not help O2 to pass the second linear algebra course. The practice of focusing on discourse rather than on thinking risks discriminating against individuals with highly creative mathematical minds.

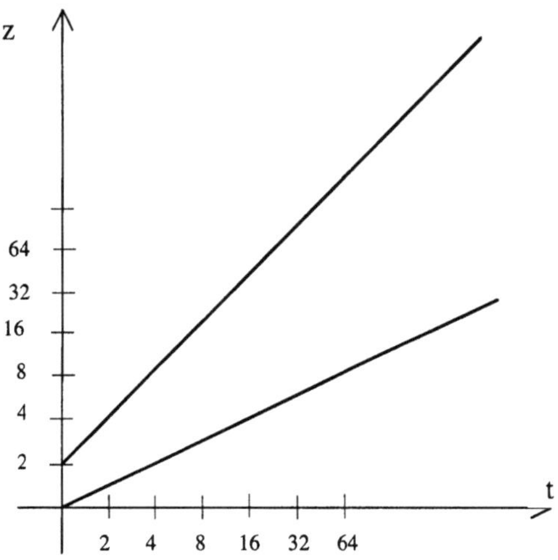

Figure 4: Which if any of the lines could represent a linear function?

3. CONCLUSION

In her 1982 thesis, Colette Laborde defined language as a code. That did not prevent her from conducting a thorough study of discursive practices in mathematical textbooks and classrooms as well as proposing didactic situations in which communication was not just a part of a didactic or experimental contract but constituted the very condition of completion of the mathematical task.

Jakobson's (1960) theory of language functions has been forgotten because he was a structuralist, and structuralism was criticised and rejected by authoritative opinions of socio-linguists such as Bourdieu. But even if Jakobson's analyses were aimed at identifying the internal structure of semiotic systems (poetic texts, cookbooks, architecture, music), his research was deeply informed by a vast knowledge of the historical and cultural contexts in which these systems were born. I hope to have hinted at ways in which Jakobson's theory can inspire us in interpreting instances of communication in our research.

Looking at language from the perspective of theories of representation systems, one may fail to take into account the historical, social, and cultural contexts of uses of these systems, but this study may still lead to valuable insights into the specificities of mathematical discourses, as found, for example, in Duval's work.

"Cognitivist" and "mentalist' approaches may be similarly criticised for their socio-cultural blindness and inability to explain the processes of teaching and learning in actual schools and classrooms. But they are also focused on the specificity of mathematics and therefore can be useful in helping teachers to plan what they are going to discourse about in their classrooms, and prepare them to better understand and capitalise on students' often awkwardly worded contributions to this discourse.

In his often-cited paper, Thurston (1994) mentions so many different characteristics of understanding and communicating mathematics that mathematics educators of all origins could find arguments to support their theses. DA could even find arguments to support its claim that mathematical thinking is a collective enterprise. The point is, however, that for Thurston, mathematical thinking is both a collective and an individual endeavour, which is obviously the only reasonable position one can take in mathematics education.

As mentioned in the introduction, communication is a problem in mathematics education. Communication is especially a problem in mathematics; it is a problem for mathematicians themselves (see Thurston's, 1994, insightful comments on this point). By equating thinking and communicating, the problem can be eliminated but not solved. That is why DA alone is not a sufficient theory for mathematics education.

Mathematics education has learned a lot from other disciplines. But if that knowledge does not take into account the specificity of mathematics, it remains only an ideology. Let me refer at this point to Bourdieu, whose popularity among mathematics educators has considerably increased in the last few years. In 1994, he pretty much told his admirers to stop trivialising his work and go back to their own fields and examine the practices that are specific to them. To claim, he said, that the scientific field is a social universe like other fields is "all but an astounding discovery" (p. 96); each field is regulated by its own specific social and epistemological laws, and these laws need to be identified and studied. Let us do it. Otherwise, in the fervour of studying discourse, we shall discourse the mathematics away. There are ways of not only studying but also conceptualising communication in the mathematics classroom with a focus on its mathematical content that do not lose sight of its social and cultural dimension. That is obvious, I hope, from the contributions to this volume.

REFERENCES

Adler, J. (2001). *Teaching mathematics in multilingual classrooms.* Dortrecht: Kluwer.

Arsac, G., Balacheff, N. & Mante, M. (1992). Teacher's role and reproducibility of didactical situations. *Educational Studies in Mathematics, 23*(1), 5–30.

Balacheff, N. (1986). Cognitive versus situational analysis of problem solving behaviours. *For the Learning of Mathematics, 6*(3), 10–12.

Bandura, A. (1980). *L'apprentissage social.* Bruxelles: Mardaga.

Bell, E. T. (1956). Invariant twins, Cayley and Sylvester. In J. R. Newman (Ed.), *The world of mathematics Volume 1* (pp. 341–365). New York: Simon and Schuster

Bellemain, F. & Capponi, B. (1992). Spécificité de l'organisation d'une séquence d'enseignement lors de l'utilisation de l'ordinateur. *Educational Studies in Mathematics, 23*(1), 59–98.

Benveniste, E. (1966). *Problèmes de linguistique générale, 1.* Paris: Gallimard.

Benveniste, E. (1974). *Problèmes de linguistique générale, 2.* Paris: Gallimard.

Bessot, A. & Ridgway, J. (2000). *Education for mathematics in the workplace.* Dordrecht: Kluwer.

Bingham, S. G. (Ed.) (1994). *Conceptualizing sexual harassment as discursive practice.* Westport (USA): Praegers.

Blaye, A. (1989). Interactions sociales et constructions cognitives: présentation et critique de la thèse du conflit socio-cognitif'. In N. Bednarz & C. Garnier (Eds), *Construction des savoirs, obstacles et conflits* (pp. 341–365). Ottawa (Canada): Les éditions Agence d'Arc.

Boero, P., Pedemonte, B. & Robotti, E. (1997). Approaching theoretical knowledge through voices and echoes: A Vygotskian perspective. In E. Pehkonen (Ed.), *Proceedings of the 21st Conference of the International Group for the Psychology of Mathematics Education* (Vol. 2, pp. 81–88). University of Helsinki - Lahti Research and Training Center.

Boero, P., Pedemonte, B., Robotti, E. & Chiappini, G. (1998). The "Voices and Echoes Game" and the interiorization of crucial aspects of theoretical knowledge in a Vygotskian perspective: Ongoing research. In A. Olivier & K. Newstead (Eds.), *Proceedings of the 22nd Conference of the International Group for the Psychology of Mathematics Education* (Vol. 2, pp. 120–127). *Stellenbosch (South Africa): Stellenbosh University.*

Bourdieu, P. (1980). *Le sens pratique.* Paris: Editions de Minuit.

Bourdieu, P. (1994). *Raisons pratiques: Sur la théorie de l'action.* Paris: Editions du Seuil.

Bourdieu, P. (2001). *Langage et pouvoir symbolique.* Paris: Editions du Seuil. (Revised version of *Ce que parler veut dire*, 1982, éditions Fayard).

Bruner, J. S. (1974). From communication to language: A psychological perspective. *Cognition, 3*(1), 255–287.

Chevallard, Y. (1999). L'analyse des pratiques enseignantes en théorie anthropologique du didactique. *Recherches en Didactique des Mathématiques, 19*(2), 221–266.

Chilver, P. (1982). *Learning and language in the classroom: Discursive talking and writing across the curriculum.* Oxford: Pergamon Press.

Clark, E. V. (1978). Awareness of language: Some evidence from what children say and do. In A. Sinclair, R. J. Jarvella & W. J. M. Levelt (Eds), *The child's conception of language* (pp. 17–43). Berlin: Springer.

Cobb, P., Boufi, A., McClain, K. & Whitenack, J. L. (1997). Reflective discourse and collective reflection. *Journal for Research in Mathematics Education, 28* (3), 258–277.

Corballis, M. C. (1997). Mental rotation and the right hemisphere. *Brain and Language, 57*(1), 100–121.

Coulange, L. (2001). Enseigner les systèmes d'équations en troisième: Une étude économique et écologique. *Recherches en Didactique des Mathématiques, 21*(3), 305–354.

Davydov, V. V. (1990). *Types of generalization in instruction: Logical and psychological problems in the structuring of school curricula* (J. Kilpatrick, Ed. & Joan Teller, Trans.). Reston (USA): National Council of Teachers of Mathematics.

Davydov, V. V. (1997). Lev Vygotsky and educational psychology. In *L. S. Vygotsky, Educational Psychology.* Boca Raton (USA): Lucie Press. (Original publication 1991)

Doise, W. & Mugny, G. (1981) *Le développement social de l'intelligence.* Paris: Interéditions.

Dorier, J.- L., Robert, A., Robinet, J. & Rogalski, M. (2000). The obstacle of formalism in linear algebra. In J.- L. Dorier (Ed.), *On the teaching of linear algebra* (pp. 85–124). Dordrecht: Kluwer.

Dryzek, J. S. (1990). *Discursive democracy: Politics, policy, and political science.* Cambridge (USA): Cambridge University Press.

Duval, R. (1995). *Sémiosis et la pensée humaine. Registres sémiotiques et apprentissages intellectuals.* Berne: Peter Lang.

Edwards, D. & Potter, J. (1992) *Discursive psychology.* Newbury Park (USA): Sage.

Gillet, G. 81999). *The mind and its discontents: An essay in discursive psychiatry.* Oxford: Oxford University Press.

Grize, J.- B. (1982). *De la logique à l'argumentation.* Paris: Droz.

Hadamard, J. (1945). *An essay on the psychology of invention in the mathematical field.* New York: Dover.

Harré, R. & Gillet, G. (1995). *The discursive mind.* Thousand Oaks (USA): Sage.

Hoyles, C. (2001). From describing to designing mathematical activity: The next step in developing a social approach to research in mathematics education? *Educational Studies in Mathematics, 46*, 273–286.
Jakobson, R. (1960). Linguistics and poetics. In T. Sebeok (Ed.), *Style in language* (pp. 350–377). Cambridge (USA): MIT Press.
Janvier, C. (Ed.) (1986). *Problems of representation in the teaching and learning of mathematics*, Hillsdale (USA): Lawrence Erlbaum Associates.
Kieran, C., Forman, E. & Sfard, A. (Eds) (2001). Bridging the individual and the social: Discursive approaches to research in mathematics education. *Educational Studies in Mathematics, 46*(1–3), 13–271.
Laborde, C. (1982). *Langue naturelle et ecriture symbolique: Deux codes en interaction dans l'enseignement mathématique*. Grenoble (France), Université Scientifique et Médicale and Institut National Polytechnique de Grenoble, Ph.D. dissertation.
Laborde, C. (1992). Guest Editorial. *Educational Studies in Mathematics, 23*(1), 1–4.
Laborde, C. (Ed.) (1992) Classroom dynamics. *Educational Studies in Mathematics, 23*(1).
Lacan, J. (1966). *Ecrits*. Paris: Seuil.
Lotman, J. M. (1999). *Kultura i eksplozja*. Warsaw: Panstwowy Instytut Wydawniczy. (Trans. from Russian, *Kultura i vzriv*, 1998, by Boguslaw Zylko).
Lotman, Y. M. (1990). *Universe of the mind: A semiotic theory of culture*. Bloomington (USA): Indiana University Press.
Mugny, G., Doise, W. & Perret-Clermont, A. N. (1975-1976). Conflit de centrations et progrès cognitifs. *Bulletin de Psychologie, 29*, 199–204.
Parmentier, R. J. (1985). Semiotic mediation: Ancestral genealogy and final interpretant. In E. Mertz & R. J. Parmentier (Eds), *Semiotic mediation: Sociocultural and psychological perspectives* (pp. 359–385). Orlando (USA): Academic Press.
Perret-Clermont, A. N. (1979). *La construction de l'intelligence dans l'interaction sociale*. Berne: Peter Lang.
Rossen, M., Klima, E. S., Bellugi, U., Bihrle, A. & Jones, W. (1996). Interaction between language and cognition: evidence from Williams syndrome. In J. H. Beichtman, N. J. Cohen, M. M. Konstantareas & R. Taennock (Eds.), *Language, learning and behavior disorders: Developmental, biological and clinical perspectives* (pp. 367–392). Cambridge: Cambridge University Press.
Salin, M.- H. (2002). Les pratiques ostensives dans l'enseignement des mathématiques comme objet d'analyse du travail du professeur. In P. Venturini, C. Amade-Escot & A. Terrisse (Eds.), *Etude des pratiques effectives: L'approche des didactiques* (pp. 71–82). Grenoble: La Pensée Sauvage.
Sfard, A., Nesher, P., Streefland, L., Cobb, P. & Mason, J. (1998). Learning mathematics through conversation: Is it as good as they say? *For the Learning of Mathematics, 18*(1), 41–51.
Sierpinska, A. (1996). Whither mathematics education? In C. Alsina, J. M. Alvarez, M. Niss, A. Pérez & A. Sfard (Eds). *Proceedings of the 8th International Congress on Mathematics Education* (pp. 21–46). Seville: S.A.E.M. Thales.
Sierpinska, A. (1994). *Understanding in mathematics*. London: Falmer Press.
Sierpinska, A. (2000). On some aspects of students' thinking in linear algebra. In J.- L. Dorier (Ed.), *On the teaching of linear algebra* (pp. 209–245). Dordrecht: Kluwer.
Sierpinska, A. (2002). Review of Adler, J. (2001), *Teaching mathematics in multilingual classrooms. Zentralblatt für Didaktik der Mathematik, 34*(3), 98–103.
P_2Richard, P. R. & Sierpinska, A. (Forthcoming). Etude fonctionnelle-structurelle de deux extraits de manuels anciens de géométrie. In G. Lemoyne (Ed.), *Revue des Sciences de l'Éducation, 30*(1). (Numéro Special, Le langage dans l'enseignement et l'apprentissage des mathématiques: complexité et diversité des cadres d'étude.)
Thurston, W. (1994). Proof and progress in mathematics. *Bulletin of the American Mathematical Society, 30*(7), 161–177.
Venturini, P., Amade-Escot, C. & Terrisse, A. (Eds) (2002). *Etude des pratiques effectives: L'approche des didactiques*. Grenoble: La Pensée Sauvage éditions.
Wertsch, J. V. (1991). *Voices of the mind, A sociocultural approach to mediated action*. Cambridge (USA): Harvard University Press.

MICHAEL OTTE

MEANING AND MATHEMATICS

All our intelligible relations to reality are mediated by symbols rather than being direct and objective. This thesis is uncontroversial with respect to certain types of knowledge. We cannot in principle know, for example, the interior of the atom in the way we know an apple in our garden or in the manner we know each other as we converse. What is claimed here, however, is that our knowledge of apples or of ourselves also depends on the ways we represent these things in our head or elsewhere. Every perception contains elements of interpretation as well as of generalization (because a symbol, a proposition, for example, is a general). All knowledge is thus in a certain sense indirect knowledge, being a function of the symbols and representations we use.

A symbol has meaning, but does not exist like a concrete thing, because it is a general, a type, not a token. A thing in contrast exists but has per se no meaning at all. Symbolization thus amounts to generalization. In interpretation we have to see something particular as a general. A sign is a sign of something to somebody, and as such it is also a mediation between the subjective and objective. A sign is a process as it functions to make relations effective. And as a process it is exposed to the paradoxes of movement, which Zeno already has exhibited by means of his thought-experiment about the race between Achilles and the Tortoise. This then is the "first" problem of meaning, the problem of whole and part. The second problem results from the relation between the particular and general already mentioned.

Thom (1973) in his invited lecture to the Second International Congress on Mathematics Education in Exeter in 1972, put the problem of meaning in central place. "The real problem which confronts mathematics teaching is not that of rigor, but the problem of the development of 'meaning', of the 'existence' of mathematical objects" (p. 202). And Bruner (1969) asks in a similar vein, "What do we say to a young child, asking if concepts like force or pressure really exist?" Thom, and Bruner as well, intended to draw attention to the fact that we cannot develop our cognitive activities if we do not believe in the reality of our intuitions, and that these intuitions or mental states nevertheless may be treacherous and without objective validity or reference. Subjective meaningfulness and objective validity may not coincide. To the contrary, our first ideas as a rule are nearly always false, or inappropriate.

And what is worse, intuitions are not auto-corrective. Our intuitions as such are like a conglomerate of Leibnizean Monads, each of which represents the world out of its own particular perspective. Intuitive knowledge is not discursive, or so it seems. In intuition everything appears with the same convincing clarity that distinguishes our clearest visual perceptions. The experience of the untruth of our intuitive insights only comes by way of a second intuition. If this should not lead to an aporia—intuitions do not admit questions of validity—we must dynamise our intuitions by applying them. One must draw conclusions or search to continuously transform ones intuitions or perspectives into one another. Thus intuitions must be embedded into processes of reasoning and experimenting. "The very supreme commandment of sentiment is that man [...] should become welded into the universal continuum, which is what true reasoning consists in", Peirce wrote (1958 I, par. 673). The continuum "becomes the true universal" (Eisele, 1976, p. 211).

On the other hand do these processes of reasoning and calculation not get started without a motivating idea or engaging intuition. This is what Thom and Bruner seem to have in mind. One might neither be able to pose a question or to establish the relevant mathematical equations, nor to solve them afterwards, if one is not guided by some heuristic ideas or some motivating belief in the relevance of the matter at hand. And with respect to the wider concerns of mathematical education, it seems relevant to observe that living in a so called knowledge society, as we do today, leads to the experience that knowledge claims become overwhelmingly numerous and controversial and begin to form an inconsistent and chaotic universe, which has no clear frontiers towards ignorance. The world cannot be considered independently from our modes of being in the world and vice versa.

Let us finally give some overview of the chapter. First, we outline Kant's epistemology and its consequences. Then some meaning problems are treated from the point of view of formal philosophy of mathematics, as developed by Frege and Russell. Finally, we give an exposition of Peirce's theory of meaning.

1. MODES OF KNOWING

The consideration of knowledge as an activity and as a process leads to abandoning the classical dualism of the internal vs. external, replacing it by differentiations within the system of (semiotic) activity or by distinctions between ways of knowing and faculties of the mind. Only that which is effective upon the active system can have an epistemic impact. This transformation of point of view has been accomplished by Kant already, in response to the quarrel between idealists and materialists.

Kant's greatest merits lay, according to Peirce "in his sharp discrimination of the intuitive and the discursive processes of the mind. The distinction itself is not only familiar to everybody but it had long played a part in philosophy. Nevertheless, it is on such obvious distinctions that the greater systems have been founded, and [Kant] saw far more clearly than any predecessor had done the whole philosophical import of this distinction. This was what emancipated him from Leibnizeanism, and at the same time turned him against sensationalism. It was also what enabled him to see that no general description of existence is possible, which is perhaps the most valuable proposition that the Critic contains. But he drew too hard a line between the operations of observation and of ratiocination" (1958 I, par. 35).

We shall touch on all these aspects mentioned by Peirce in due turn. Russell was a "sensationalist", in the sense intended by Peirce. Leibniz in contrast was an idealist. Kant obviously tried to avoid these extremes, but "drew too hard a line between" the types of knowing. Let us be clear about the fact that Kant's distinction is an absolute consequence of his endeavor to eliminate the classical internal/external distinction. But to really accomplish this task one has to see the relationship between subject and object as the essential reality, rather than to begin considerations with the relata, the subject as such or the object as such. Semiotics or the theory of symbolism does exactly that. If there existed a direct and unmediated access to the objects of knowledge, then this would also exist in a quasi-automatic or mechanical form (see also Otte, 1991). Kant's distinction between the discursive and intuitive processes of thought would thus not make sense any more. If things were identical with their (mental) images there were no need for an epistemology or a learning theory.

But are not our perceptions direct and unmediated? Perhaps, but empirical perception does not per se provide knowledge, says Kant. For Kant, when we operate in the mode of knowing, all our sensibility is laden with forms of intuition and with certain concepts. This implies by means of contrast that reality as such, or the thing in itself, is cognitively ineffable and is completely unspecifiable. If we humans possessed, Kant says, the power of intellectual intuition, as we do not, the thing in itself would be "noumenal" for us, that is, a being of reason, and we would be able to know it without the constitutive contribution of the world of phenomena. Kant's great insight consists in pointing out that complementarity between things and sensible appearances of those, dividing the realm of the objective into complementary regions.

Kant writes:

> The possibility of a thing can never be proved merely from the fact that its concept is not self-contradictory, but only through its being supported by some corresponding intuition. [...] If, therefore, we should attempt to apply the categories to objects, which are not viewed as being appearances, we should have to postulate an intuition other than the sensible, and the object would thus be a noumenon in the positive sense. Since, however, such a type of intuition, intellectual intuition, forms no part whatsoever of our faculty of knowledge, it follows that the employment of the categories can never extend further than to the objects of experience. [...]
>
> If we abandon the senses, how shall we make it conceivable that our categories [...] should still continue to signify something, since for their relation to any object more must be given than merely the unity of thought—namely, in addition, a possible intuition, to which they may be applied. [...] For we cannot in the least represent to ourselves the possibility of an understanding which should know its object, not discursively through categories, but intuitively in a non-sensible intuition" (Kant, B308-B312).

Peirce made the following comment with respect to Kant's *Critique of Pure Reason*:

> The first step of Kant's thought [...] is to recognize that all our knowledge is, and forever must be, relative to human experience and to the nature of the human mind. That conception being well digested, the second moment of the reasoning becomes evident, namely, that as soon as it has been shown concerning any conception that it is essentially involved in the very forms of logic or other forms of knowing, from that moment there can no longer be any rational hesitation about fully accepting that conception as valid for the universe of our possible experience. To repeat an example I have given before, you look at an object and say 'That is red'. I ask you how you prove that. You tell me you see it. Yes, you see something; but you do not see that it is red; because that it is red is a proposition; and you do not see a proposition. What you see is an image and has no resemblance to a proposition, and there is no logic in saying that your proposition is proved by the image. [...] At this point, the idealist appears before the tribunal of your reason with the suggestion that since these metaphysical conceptions, that repose upon their being involved in the forms of logic, are only valid for experience and since all our knowledge is relative to the human mind, they are not valid for things as they objectively are; and since the conception of existence is preeminently a conception of that description, it is a mere fairy tale to say that outward objects exist, the only objects of possible experience being our own ideas. Hereupon comes the third moment of Kant's thought, [...] It is really a most luminous and central element of Kant's thought. I may say that it is the very sun round which all the rest revolves. This third moment consists in the flat denial that the metaphysical conceptions do not apply to things in themselves. Kant never said that. What he said is that these conceptions do not apply beyond the limits of possible experience. But we have direct experience of things in themselves. Nothing can be more completely false than that we can experience only our own ideas. That is indeed without exaggeration the very epitome of all falsity. Our knowledge of things in themselves is entirely relative, it is true; but all experience and all knowledge is knowledge of that which is, independently of being represented. Even lies invariably contain this much truth, that they represent themselves to be referring to something whose mode of being is independent of its being represented. (Peirce, 1958 VI, par. 95)

That all our knowledge is related to experience and intuition does not imply thus that it is merely subjective. Only the forms and categories by which we make explicit what we know are characteristics of the (transcendental) subject. All our thinking is by signs, but a sign is co-determined by its object and is not just of our arbitrary making. Cassirer (1962) also links Kant's rejection of the notion of an intellectual intuition, by which humans could cognise things in themselves directly and absolutely, with the symbolic nature of human knowledge. He writes:

> In his *Critique of Judgment,* Kant raises the question whether it is possible to discover a general criterion by which we may describe the fundamental structure of the human intellect and distinguish this structure from all other possible modes of knowing. After a penetrating analysis, Kant is led to the conclusion that such a criterion is to be sought in the character of human

> knowledge, which is such that the understanding is under the necessity of making a sharp distinction between the reality and the possibility of things. [...] Human knowledge is by its very nature symbolic knowledge. It is this feature, which characterizes both its strength and its limitations. And for symbolic thought it is indispensable to make a clear distinction between *real* and *possible*, between actual and ideal things. A symbol has no actual existence as a part of the physical world; it has a meaning" (Cassirer, 1962, p. 86).

There is, however, not only a contrast here between the factual and the merely possible, but also a connection. Otherwise Kant's notion of "objective possibility" (in contrast to the merely logically possible) would be disregarded. Thus the distinction of symbols and things seems fundamental because meanings are, on the one hand, universals, but must be given by means of token or objects, on the other hand. Meanings are ideal entities, like (natural) laws and, like the latter, they must be verified by specifying their intended interpretations or applications, rather than just giving a linguistic circumscription of them.

Kant pointed out that continuity and coherence of our cognitions are of fundamental importance. Continuity results from the fact that our perceptions or intuitions are of a continuous nature. Kant says that the "principle of perception" consists in the fact that in all appearances the objective has degrees or nuances (Kant, B207). "All appearances therefore are continuous quantities, be it, according to intuition, as extensive quantities, be it, according to perception, as intensive quantities" (B212). It is true that we can grasp only that which has meaning for us and that meaning depends on continuity or relationship. What turns an individual observation or mental event into thought, what provides a particular statement with meaning, is, on the one hand, its connection with other such events or statements. There is no knowledge without form or structure. Kant pointed out that this form cannot be provided by logic and language alone, but must be based on forms of intuition. Concepts do not apply to things in themselves but rather to intuitions or mental representations of objects. Or, as Kant had stated it, coherence and logical consistency without sensuality and intuition do not give objective possibility of thought, which is rather determined by the constraints of human experience.

Following Cassirers affirmations one is lead to ask how symbols and things might possible come together—or stated in Kantian terms, how concepts and intuitions come together—thereby constituting objective knowledge. Now this question must be answered in genetic terms. This was the reason, which led Kant to transform the duality of the external and internal establishing human activity as the essence of the mediation between subject and object of knowledge. But Kant gave activity a too much subjective turn. He did not see that we might have objective experience of proximity and continuity, rather than synthesising the appearances of isolated distinct "things in themselves" according to the forms of pure intuition. We cannot rationally grasp things in themselves, as Kant had affirmed and Peirce consented—see the citations given—, but we can perceive relations of contiguity.

"Kant gives the erroneous view", writes Peirce, "that ideas are presented separated and then thought together by the mind. This is his doctrine that a mental synthesis precedes every analysis. What really happens is that something is presented which in itself has no parts, but which nevertheless is analyzed by the mind, that is to say, its having parts consists in this, that the mind afterward recognizes those parts in it. [...] When, having thus separated them, we think over them, we are carried in spite of ourselves from one thought to another, and therein lies the first real synthesis. An earlier synthesis than that is a fiction" (Peirce, 1958 I, par. 384).

The problem results from Kant' failure to acknowledge the reality of relations and continua and from the fact that a logic of relations had not yet been developed in his times. Kant says that we do not have axioms in arithmetic, because statements like "7 + 5 = 12" have nothing general to themselves (Kant, B206). Number symbols seem to be proper names of objects and the rules for their manipulation are to be constructed only afterwards. According to Kant, a theorem like "7 + 5 = 12" is to be considered synthetical knowledge, because "the conception of a sum of 7 and 5 contains nothing but the uniting of the two numbers into one, whereby it cannot at all be cogitated what this single number is which embraces both. The conception of twelve is by no means already obtained by merely cogitating the union of 7 and 5; [...] One must go beyond these concepts, and have recourse to an intuition" (Kant, B15–16).

The dominant philosophy of mathematics considers this distinction or contrast between the factual and the general or between the existent and the merely possible as being of absolute nature. Russell, for instance, endorsing an absolutely realistic conception of logic and mathematics, draws a firm distinction between names or indices on the one hand and descriptions on the other, and believes that names have no place in pure logic and mathematics. Logic or mathematics constructs symbols, which help to grasp the most general aspects of concrete reality. But neither mathematics nor common knowledge bother too much about the ontological status of their creations. Numbers are interpreted as names of sets of sets, for instance, but there seems little consensus among logicians and philosophers about the question as to what sets really are. It is a fact, according to Russell, that our descriptions or words in general are related to other words or descriptions only, rather than to what these words mean. "All statements about unicorns are really about the word 'unicorn'", says Russell (2000, p. 68) and the same applies to nearly everything we are speaking about, namely to everything which we do not directly perceive. Is thus any proposition, which does not speak about our actual percepts meaningless? (see Russell, 2000, p. 632).

One consequence of Russell's absolute separation or distinction between direct and indirect knowledge or between indices and descriptions, is that there is no continuity in our thinking. We have, for instance, believes Russell, "no knowledge of the past" (Russell, 2000, p. 69). And we therefore do not live in any concrete world, but seem to have been fallen from the concrete into the abstract. All our cognitive activities seem to be some sort of linguistic "make-believe games".

One might claim, in conclusion that neither Kant nor Russell are right and that mathematical statements represent neither purely analytical nor synthetic knowledge.

We may in fact characterise the historical development of mathematical thought by saying that this development is nothing but an unending endeavour to reformulate the "principle of continuity", which is the fundamental principle of mathematical generalisation. Mathematics is relational thinking of which algebra furnishes the first clear expression. Relational thinking begins already with Eudoxus' theory of proportions as presented in Book V of Euclid. Mathematicians know that no theoretical entity is really defined without giving criteria for its identity. Definition 5 of Book V of Euclid presents Eudoxus' definition of equality between two relations: ${}^{a}/_{b} = {}^{c}/_{d}$. This definition in a sense uses the principle of continuity, as Dedekind's definition of real number was to reveal much later.

Continuity can only be experienced and this experience is best to be gained through continuous efforts of representing and re-representing over and again what appear to be the same mathematical facts. The difficulties of pupils with respect to this activity are very well known. It is reported, for example, that students have great difficulties with the distinction between rational and irrational numbers. 1/3 is sometimes considered "more" rational than 0,33333... by pupils who have little experience with the variability of symbolisation (see Toerner, 2003). And Felix Klein (1924, p. 36) made the observation already that physicists or astronomers considered 2/7 as rational but not 2021/7053.

We take it for granted therefore that meanings have a dual nature in as much as the reality of our meanings depends as much on the reality of continua or laws as it requires concrete applications and firm relations to matters of fact.

1.1 An example where the object has been lost: Incommensurability

Mathematical thinking, like thinking in general, needs a content or object to become real thinking, rather than mere playing with symbols. By object we mean anything to which the symbols refer and which is not just a description. The object of thought thus is something different from the sign or from any representation of it. Kant believed that the object of thought is given to us by intuition. We need not quarrel about intuition here. We should recognise, however, that Kant had pointed to a very important cognitive fact. Meaning is not just a linguistic or conceptual matter. The difference with a narrow empiricism and verficationism becomes salient in the following two examples of explaining the nature of incommensurability.

> *Commensurable line segments*
>
> In comparing the magnitudes of two line segments a and b, it may happen that a is contained in b an exact integral number r of times. In this case we can express the measure of the segment b in terms of that of a by saying that the length of b is r times that of a. Or it may turn out that while no integral multiple of a equals b, we can divide a into, say, n equal segments, each of length a/n, such that some integral multiple m of the segment a/n is equal to b:
>
> $$b = \frac{m}{n}a$$

> When an equation of the form above holds we say that the two segments a and b are *commensurable,* since they have as a common measure the segment a/n which goes n times into a and m times into b. (Courant & Robbins, 1941, p. 58)

> *Commensurable line segments*
>
> It is said that two line segments are commensurable if they have a common measure. What does it mean to have a common measure? Let us assume that one line segment is 3 cm long and another, 9 cm. The two line segments are commensurable: The common measure is 3 cm. It fits once into the first line segment and exactly three times into the second. Let us assume that one line segment is 6 cm long and another, 10 cm. These, too, are commensurable. Their common measure is 2 cm: It fits three times into the first line segment, and five times into the second. Even for two line segments of length, say, 1.67 cm and 4.31 cm, it is easy to find a common measure: 0.01 cm. It fits 167 times into the first line segment and 431 times into the second. What do these examples tell us? Two line segments are commensurable if one line segment (or a fraction of it) is contained within the other without remainder. (Thun & Götz, 1976, p. 47, my translation)

The second quotation above was written by two psychologists who wanted to "improve the original mathematical text" by Courant and Robbins. "Avoiding variables, formulae, and diagrams" was noted as a typical feature of improvement (and the revision also omitted a geometrical diagram that was in the original). Indeed, the revision has its merits from the perspective of "pure" readability, which is conceived of as being neutral with regard to a cognitive use of the text. The second text seems clear and straightforward.

On the other hand, the revisers do not seem to have realised that the mathematical subject matter itself has in a way disappeared after the variables and diagrams have been eliminated. If one replaces the relations between line segments by relations between decimal numbers from the very outset, one of course always has a common measure. The object of the original text does not simply consist of a defining circumscription of commensurability; it was occasioned by the problem of incommensurability, which has continued to cause astonishment, speculation, and contemplation since antiquity. This is the question at issue mathematically, and not the verification that 1.67 is a rational number, as the second text leads one to believe (see Otte, 1994). One might even suppose, reading this text, that all numbers are rational and that there is no incommensurability. If somebody wanted to use the arithmetical language, he or she would have to furnish at least one example of an irrational number. This is commonly done, by characterising the *class* of rational numbers in a formal way (as the periodic decimals, for instance).

The subject matter in question, namely incommensurability, appears then as the unknown, or at least, as the territory not yet described and mapped. The irrational is characterised in merely negative terms, that is, as that which is not rational, and it stretches beyond the borderlines of rationality. It is not given, but is rather the challenge, it is the territory yet to be conquered. This is the way Peirce has defined the continuum. "as that every part of which can be divided into any multitude of parts whatsoever" (1958 III, par. 569). And further: "A true continuum is something whose possibilities of determination no multitude of individuals can exhaust" (1958 VI, par. 170).

If one wants to prove, however, that some X is not a rational number, for example, or is not a constructible number, this X must be characterised positively to begin with, and must be given as a particular element of a positively delimited universe of discourse. This endeavour requires something more than just decimal arithmetics.

2. MEANING AND REFERENCE

A representational "equation" $A = B$ is commonly interpreted as saying that A and B are different intensions of the same extension. In his famous essay *"Über Sinn und Bedeutung"* ["On sense and reference"], Frege (1892) gives some examples from elementary geometry. He writes:

> Let *a, b, c* be the lines connecting the vertices of a triangle with the midpoints of the opposite sides. The point of intersection of *a* and *b* is then the same as the point of intersection of *b* and *c*. So we have different designations for the same point, and these names ("point of intersection of *a* and *b*", "point of intersection of *b* and *c*") likewise indicate the mode of presentation and hence the statement contains actual knowledge.

Very often one has to find out yet whether A and B represent the same thing. The mathematician's task, indeed, "can be described as the transformation of facts into consequences" (de Gandt, 1995, p. 121). The mathematician accomplishes that task by gradually transforming the initial facts or givens until something really new appears. A series of equalities thus leads to an inequality: $A = B$ and $B = C$, but A is unequal to C. Hilbert had characterised the continuum in this way, but had it thought not fit for mathematics because of the fact that the relations are not transitive. A set theoretical model of it thus should substitute the continuum (Hilbert, 1992).

Mathematical deduction, however, is perfectly characterised by this continuity as it could be understood as the interpretation or translation of one sign into another (see also Peirce, 1958 V, par. 53). What guides these translations cannot be just logic, because formal logic can only operate within one fixed formal context. Deductive reasoning seems to unfold the intensions of the theoretical terms as fixed in the premises and axioms. This process is logically constrained but is not always totally determined or preprogrammed, because in the more complicated cases a generalisation might be required and the creation of new concepts.

Peirce affirms that in these cases a "theorematic deduction" occurs, which "performs an ingenious experiment upon the diagram, [that is, on the image of the premisses] and by the observation of the diagram, so modified, ascertains the truth of the conclusion" (1958 II, par. 267). This modification depends on observation, perception and the abductive introduction of a new idea according to which the diagram is then modified in order to render the conclusion more or less obvious. One could argue that these new ideas are subject to a principle of continuity. The deductive process is split up into so many small steps that the conclusions that lead from one step to the next in the argument become obvious and perceivable. Each step results in a statement of the form $A = B$. This formulation is shown in more detail in Otte (2002), which discusses the use of Cabri geometry. What guides the subject in her/his reasoning is some heuristic idea, which determines what the object of discourse is to be like or what counts as a legitimate transformation from A into B. It is important to note that there is nothing in the world that will a priori guarantee the success of such an activity.

One might object that we are here confusing mathematics as such with the manner humans acquire mathematical knowledge or prove mathematical truths. Such a criticism had been brought forward, for example, against Kant's conception of mathematical knowledge by Bolzano already. Bolzano was a mathematical Platonist. But Platonism cannot answer the question how we can have access to the Platonic heaven of ideas in themselves and it is therefore not very popular nowadays. Besides it might be claimed that there is no mathematics as such. The extensions of terms, about which Frege speaks, for example, are in general not given as such, but are theoretical entities to be created. In the examples of mathematical entities or in case of concepts like energy (of which, e.g., heat and motion are different representations) or like the electro-magnetic field, we do not deal with an empirical object but rather with a universal object or a objective relation.

Structuralism conceives of such entities as determined by an axiomatic system. But whence do the axioms come from? And how could we change or generalise given axiomatical systems? The axioms represent the characteristics of the objects of our universe of discourse and they seem to determine it. Even if this were admitted the question remains how do we choose or change our universe of discourse. George Boole in 1853 already indicated the problem:

> If the universe of discourse is the actual universe of things, which it always is when the words are taken in their literal meaning, then by men, for example, we mean all men that exist; but if the universe of discourse is limited by any antecedent implied understanding, then it is of men under the limitation thus introduced that we speak. [...] The operation which we really perform is one of selection according to a prescribed principle or idea. (Boole, 1958, pp. 42–43)

This "implied understanding" or heuristic idea need not be completely explicit and thus is not to be identified with a fixed axiomatic system, although we might begin with such a system, when trying to prove a mathematical proposition, for example. But mathematical proofs, in their more complicated cases, when they do not boil down to a straightforward verification, cannot be accomplished without some generalisation. The establishment of equations such as $X = Y$ or $A = B$ thus depends on generalisation and abductive reasoning. Ideal objects or universal relationships have to be postulated after all. Conversely, the essential features of an act of imaginative creation may be summarised by stating that they consist in seeing an A as a B: $A = B$, or "all A are B", or "A represents B", and so on. This creative act may be attempted in the service of solving a specific problem, or it may be intended to explain some fact of the matter. A mathematical explanation of a certain fact A is really nothing but the furnishing of a second perspective B on that fact: $A = B$.

Mathematical activity consists to a large part in proving. The perspective of argument and proof inevitably turns mathematics into a collection of statements. Knowledge in general according to the linguistic consensus, consists of systems of propositions and it is therefore in these in which the knower has justified true belief. Proofs enlarge this confidence as they connect propositions into longer chains of reasoning. Deductive reasoning strengthens the conviction that to be real means to be an element of a system; and by system a conceptual system is meant. It is true that we can grasp only that which has meaning for us and that meaning depends on structure or more generally, on continuity.

In Otte (2002) we have provided a new proof of Euler's theorem that certain points in the triangle are collinear by interpreting the relevant diagram in such a way as to show that Euler's theorem is a special case of a Desarguen configuration. Or say, we have accomplished the proof by claiming that the meaning of the diagram in question "really" is "such and such". We have thereby moved from the context of Euclidean geometry, in which the original theorem had been formulated, to the more general context of projective geometry. And thus we generalised the given structure.

According to the axiomatic approach, one would claim that the structures in question themselves determine objective possibility, in the sense that logical necessity determines future development. "Every system must therefore proceed in the direction in which its own consistency can be reinforced" (Beth & Piaget, 1966, p. 275). One would then have to answer the question of where the axioms come from and how they are established and, finally how mathematical generalisations or ideal objects can be justified. Piaget's genetic epistemology tries to answer those questions. Piaget (1981) believes that "the set of all possibilities is as antinomic as the set of all sets" and he thus justifies the importance of an operative approach to mathematics and of a genetic epistemology. The cognitively possible, Piaget says "is essentially an invention and creation and therefrom derives the importance of its study for epistemological constructivism" (p. 6). This invention is, however, constrained by the structures already constructed and thus actualised.

> We have no right to talk of the system of possibilities, or so it seems, and say something valid about it before it has been realized in effective operations, that is, before the possibilities have ceased to be mere possibilities. All that we can say, and verify, about relations between the possibilities and their realization in a new logico-mathematical construction is that, genetically speaking, a structure observed at a given level of development always contains more possible generalizations (for example, by raising a restriction or abstracting a new transformation etc.) than the subject perceives. [...] Can we then limit ourselves to saying that any structure, however elementary, provided that it be of a logico-mathematical nature [...] involves a whole system of possible developments, and that the novelty of later structures consists merely of actualizing some of them? This is our hypothesis. [...] But what we object to, for genetic reasons, is the transition from the possible to the real entity so long as there has been no actualization by an effective construction" (Beth & Piaget, 1966, pp. 207–301).

What Piaget and mathematical structuralism in general miss completely is the fact that without the spatial metaphor and intended applications we would neither be able to establish the consistency of the structures nor to complete them. How could one possibly establish identity or equivalence relations between actions? How could one define the inverse of an operation? This dilemma means we are back not to Kantian intuitionism but certainly to a dualistic picture of mathematical meanings.

Thom's and Bruner's formulations, as quoted above, could also be interpreted as saying that the problem of meaning and cognition or learning depends on a sense of confidence, a feeling of relevance, and an intuitive intimacy with some objective content, as though cognitive activity were like moving around within a new building or an interesting town for the first time. Even when one is unfamiliar with the specific features, one nonetheless knows what a town is, how to explore it, or how to gather information about it. In like manner, one has to experience first of all, what kind of entities theoretical concepts or objects are and how one treats or uses them. After that only, is it possible to pose particular questions of ones own and thus to learn. All knowledge, therefore, seems to have experiential or perceptual origins. Piaget does not really confront the question of the cognitive status of the possibilities that determine further development, because he does not treat the perception of continuity, which alone provides generality on the mathematical structures (see, e.g., Peirce, 1958 V, par. 150). Piaget's constructivism, Thom (1973) believes, is "hopelessly enmeshed in difficulties linked to the following problem: how can geometrical continuity arise from a discrete 'dust' of psychological states or processes?"

Thom, in a quite Peircean spirit, repeatedly emphasised, that if a person possesses any consciousness at all, it will include consciousness of time and space. Geometrical continuity is absolutely inseparable from mathematical thought. Thom, as well as Peirce, therefore claims that no new knowledge and no generalisation can be brought about without the perception of continuity and without employing icons or diagrams, together with a principle of continuity. Hence the importance of mathematics is, Peirce says "that all mathematical reasoning is diagrammatic and that all necessary reasoning is mathematical reasoning, no matter how simple it may be" (1958 V, par. 148). Diagrams are essentially icons, and icons are particularly well suited to make graspable and conceivable the possible and general rather than the actual and existent.

It should be mentioned in this context as well that psychology and psychotherapy have known for some time that icons or images are particularly well suited to strengthen what could be called a "sense of possibility", which seems indispensable to a person's mental health. But the strongest criticism Thom launches refers to the notion of space and continuity taken as a metaphor of reality itself. Reality is, however conceived of in terms of possibility or potentiality, rather then in terms of existents or collections of distinct existents, because reality is seen as offering room for ever new generalisations. "Continuity, as generality, is inherent in potentiality, which is essentially general", says Peirce (1958 VI, par. 204).

Throughout history there has been this double orientation towards the distinct and the continuous. Aristotle, although the foremost representative of classical logic—which rests on the assumption of the possibility of clear divisions and rigorous classification—had already made the observation that "nature refuses to conform to our cravings for clear lines of demarcation", and "he first suggested the limitations and dangers of classification" (Lovejoy, 1936, p. 56). Aristotle thereby became responsible for the introduction of the principle of continuity into natural history. "And the very terms and illustrations used by a hundred later writers down to Locke and Leibniz and beyond, show that they were but repeating Aristotle's expressions of this idea" (p. 58).

Ever since Aristotle, these two sides of cognition, the intuitive or figurative and the conceptual and operative, have remained opposed to one another. It is on such distinctions between the intuitive and the discursive processes of the mind, writes Peirce, "that the greater systems have been founded" (1958 I, par. 35). Mathematics seems completely analytical and based on rigorous definitions. Its dynamics, however, and the possibility of its further evolution depends on the not yet defined and little understood. Mathematics therefore has at all times tried to come to grips with continuity and has over and again given ever new and different interpretations of the principle of continuity. Now the principle of continuity was meant to give the general and the particular their complementary roles. It has at all times been intended to save mathematics from machine like determinism. Lovejoy concluded, for example, that the continuity principle had to be transformed into a temporalised form at the beginning of the 19th century with the appearance of evolutionistic theories, on the one hand, and of constructive structuralism, on the other.

One may summarise Thom's criticism by emphasising once more that meanings have a dual structure and that the referential aspect cannot be conceived of in terms of collections or sets, but rather in terms of continua or natural kinds (with respect to the latter term, see Quine, 1969). Or stated differently, the meaning of a mathematical term has two sides, an intensional one referring to the collection of conclusions to be derived from it together with the axioms of the related theory, and an extensional side, which remains largely unknown and ill defined, which however must be experienced by the active mind to gain strength and purpose. Some time ago, Castonguay (1972), following ideas that stem from remote sources, like the famous Logic of Port Royal of 1662, had sketched such a dualistic conception of meaning in mathematics. Castonguay tried to lay bare two objective components of meaning, one of which refers to objects, and which it is appropriate to name the extensional, or correspondence component of meaning; the other relating to concepts or linguistic expressions, and which it is suitable to call the intensional, or coherence component, in that it expresses how a given concept or expression coheres, or hangs together, with its fellows through relations of consequence.

I assume that the extensional component of mathematical meaning, as a result of the absence of an authentic referential pole—after all, numbers or functions are not like chairs or other concrete objects—is to be conceived of in a twofold way. On the one hand, it is understood as something that Castonguay (1972) has called the heuristic component of mathematical meaning and that denotes a source of inspiration "for the positing of relations between variously (and possibly referentially perceived mathematical concepts or entities, relations which may eventually crystallize, through more exact formulation and deductive corroboration, into objective relations of entailment between linguistically expressed concepts" (p. 3). I intend this heuristic component to be the set of all possible representations of a mathematical relation or the class of all possible applications of a mathematical structure.

The meaning of a term or a concept evolves as soon as this concept is used and applied within a variety of different objective contexts and with different tasks in mind. Students are very often advised to strictly stick to the definition of a concept, letting no intuitions of their own and no free associations of ideas interfere. This advice, although it has some justification—as concepts serve to distinguish and classify phenomena, and definitions reflect exactly this function—is one sided and essentially misguided. It is true that everything seems metaphorically related to everything else. To counteract the resulting vagueness, one tries to be precise in one's language and at the same time one has to work within objective contexts. A concept may change its meaning from one context to another, and it should do so, in order to be useful as a general or universal. The creation of concepts is the driving force behind generalisation, which in turn is the essential business of every science, including above all mathematics.

A concept is therefore to be distinguished from any of its definitions or applications. Stated in semiotic terms, this relationship between the particular and the general that makes up for an essential aspect of the problem of meaning, amounts to the following: A mathematical object, such as "number" or "function", does not exist independently of the totality of its possible representations, but it is not to be confused with any particular representation either. It is a general idea that cannot be exhausted as such by any number of its representations. It is not only general but also a means or an instrument of cognitive activity. I have elsewhere explained the resulting complementary of mathematical concepts (Otte, 1990, 1994; Otte, Steinbring & Stowasser, 1977). A mathematical proposition or problem is an objective structure that, however, has no meaning apart from its possible representations—or what amounts to the same thing, apart from the ways it could be understood. Again, I conclude that meaning is linked with continuity or variation of understanding and that language must be seen as a collective enterprise to begin with.

3. MEANING AND EXISTENCE

Let us consider again Thom's indication that meaning, or the experiencing of something as meaningful, necessarily requires the belief in its existence. Existence cannot be defined nor proved, I believe, but must be indicated or postulated. Mathematical axioms in the modern, Hilbertian sense make no existence claims with respect to the objects described by them. This characteristic stands in definite contrast to Euclidean axiomatic. One cannot establish existents by means of language nor fix the referents of linguistic terms by descriptions. "The actual world cannot be distinguished from a world of imagination by any description" (Peirce,

1958 III, par. 363). This insight had already been expressed by Kant, who said, "Being is not a real predicate" (Kant, B622). And it had let him to emphasise the distinction between the intuitive and the discursive (see the second section above). Kant's conviction is a result of his criticism of the ontological argument for the existence of God. The kernel of this argument, by the rationalists of the 17th century was to claim, that the notion of the nonexistence of God is a contradiction: God is perfect and existence is perfection, so God must exist. Rationalism depended on this argument because God's mind was supposed to provide stability, generality, and truth for human intuition and knowledge (see Hacking, 1980). The immense dynamics that resulted in the scientific revolution of the 17th century was based on bold imaginations of the infinite and fostered and expressed by new representational systems and operative ideas as their meanings. Infinite series are an example of fundamental importance: they were taken formally and primarily subject to certain arithmetical manipulations, convergence was not an essential property to begin with. The new mathematics and logic of the time was conceived of as purely constructive or synthetic. What was to be constructed, were ideas or concepts, rather than objects. The mathematical ideas were the objects themselves, capable of being represented formally and existing only in the mind (of God). Leibniz argued that in reality there are no indivisible atoms and that the simple indivisible and enduring substances, on which everything is to be based, must be spiritual entities or monads. Therefore the ideal has to found the existent, and for this God's mind was indispensable. The (natural) laws must found that, which is governed by them. An infinite arithmetical series, for example, is to be defined in terms of a law or algorithm.

Kant objected to the ontological argument, and one of his reasons was that, in his view, existence is not a property. Since to say of some x that it exists adds nothing to the concept of x, "exists" is not a predicate. Kant concluded that there are two sources of knowledge—concepts and intuitions—and he made the latter the basis of possible existence claims.

"What has been said so far is not quite right as it stands," Grayling (1997) believes, "for, in a sense, to say of tigers that they exist *does* add something; it says that the concept of a tiger has instances in reality—that is, that *there are* tigers to be met with in the world" (pp. 89–90). Kant would not deny that but demands an intuition for establishing such extra knowledge; in the case of mathematics this intuition is what Kant had termed "pure intuition", that is, "the form under which something is intuited". Concepts without intuitions remain empty, he says (Kant, B76). A statement of existence is accordingly in fact, a higher-order statement involving reference to a concept. It becomes a second-order predicate, a predicate of predicates, saying that the predicate in question has applications; or, to state it in Kantian terms, that it is "not empty".

"Existence is essentially a property of a propositional function. It means that the propositional function is true in at least one instance" (Kant, B233). McGinn (2000) describes this traditional conception as follows: "When you think that tigers exist you do not think of certain feline objects that each has the property of existence, rather you think of the property of tigerhood, that it has instances [...] The concept of an object existing simply is the concept of a property having instances" (p. 18).

Or in Russell's terms, to say that tigers exist, is to say that "x is a tiger" is sometimes true.

There are several kinds of objections against the orthodox view of existence. First, that one cannot define existence in this way: "Since the notion of instantiation must be taken to have existence built into it—it must be *existent* things that instantiate the property" (McGinn, 2000, p. 22). One thus has to postulate or establish existents in a different manner. Russell, in fact, at first assumed existence axiomatically, by means of his "axiom of infinity" (see Russell, 1970, ch. 13). In consequence of Russell's realistic conception of logic, "logical truths are truths about this world, the only world our language can speak about, albeit its most general and abstract features" (Hintikka, 1997, p. 29). What exists is thus decided by the meanings of our concepts, and this view seems to come down to an empiricist variant of Kantianism, with set theoretic intuition taking the place of Kant's "pure intuition". And as Kant's pure intuitions seem not to be too different from Leibniz' ideas, the relationship between the intuitive and operative aspects of a concept seems to pose the major problems of understanding.

It seems that existence of particular entities was not Thom's and Bruner's concern nor the real world applicability of our concepts and theoretical ideas. Heuristically or cognitively there are general ideas, like the famous general triangle of school geometry for example, that are not predicative and that cannot be determined extensionally in terms of collections of particular instantiations. There does not exist anywhere an instance of a general triangle. In mathematics, entities like the general triangle are a free variables and not collections of determinate triangles.

Such general ideas are objects of some representation in their own right; they are concrete universals, as they are sometimes called. They are "real" because the mathematician proves true theorems about them. They might better be conceived of as objective possibilities or as something that is merely potential, rather than as existent particular objects. Any symbol actualises a possibility, as was argued above. A general triangle is not completely determined by its properties. Which properties are essential to a "general triangle" depend on the context—on the activity and its goals. If the task, for instance, is to prove the theorem that the medians of a triangle intersect in one point, the triangle on which the proof is to be based can be assumed to be equilateral without loss of generality because the theorem in case is a theorem of affine geometry and any triangle is equivalent to an equilateral triangle under affine transformations. This fact considerably facilitates the proof because of such a triangle's high degree of symmetry. The truth of a proposition about the "general triangle" then means nothing but that this proposition is provable in a certain way; that is, that a certain proof scheme applies. When—while speculating about the possibilities of a proof—we analyse the premises, we often have to introduce additional hypotheses, generalising our concepts, and this activity is directed at possible objects like the general triangle. Thus the "heuristic component" of mathematical meanings becomes essential. This example shows that the problem of the relationship between the operative and intuitive side of a concept or idea, which was stated above and which bothered Leibniz and Kant defines the objectivity of cognitive activity.

Russell and Frege conceive of the universal or general in predicative terms or in terms of functions, which depend in a sense on the reality of objects or arguments. Concept and function then are to be identified by means of the axiom of extensionality. Instantiations under the extensional view establish the identity and existence of concepts or propositional functions rather than being themselves established as existents by these functions. This extensional view deprives meanings of their dynamic qualities. Two concepts could be extensionally equivalent and yet could be different and might function differently within a certain cognitive context. Thus *Sinn* or meaning must be relatively independent from reference and, secondly, reference should be construed in model theoretical terms.

It should be mentioned, however, that Frege had tried also to establish the objectivity of Sinn (meaning) in operative terms; like when one is employing terms like the "general triangle". There is to be one meaning had by an expression or term, which is grasped by those familiar with the expression's use. Frege (1892) had emphasised that *Sinn* (meaning) must be something logically objective and must therefore be sharply distinguished from ideas or mental images and feelings or any other items to be found in people's minds. The meaning of a sign "may be the common property of many and therefore is not a part or mode of the individual mind". Meanings are, however distributive categories, like natural kinds, rather than collective ones or sets; and they thus are not predicative universals or extensions of concepts. This turns it so difficult for a logician to deal with the problem of meaning. How can one, it may be asked for example, control the use of various different ideal objects?

The mathematician, in fact, uses *exists* as a predicate but uses it relative to an intended universe of discourse. Does the real number x exist that makes the equation $x^2 = -1$, or written differently, $x = -{}^1/_x$ true? If so, it must be equal to 1 or to -1, which yields $1 = -1$, a contradiction. But the mathematician enlarges his or her universe and finds a new system of numbers: the complex numbers. Truth seems to depend on consistency. But consistency is given relative to a possible world or model. "The development of the notion of model and the emergence of the idea of truth have gone largely hand in hand in our century", writes Hintikka (1997, p. 29). Neither existence nor identity can be defined absolutely; both must be stated or affirmed, which can be done only relative to some universe of discourse.

From an intensional point of view, everything that obeys certain axioms is to be called a (complex) number. The consistency of that assumption needs to be confirmed by a suitable generalisation or by means of a model. As long as the imaginary number was admitted to arithmetic as a calculative symbol only, it produced the most horrible confusion (Nahin, 1998). Only after Gauss gave a relational interpretation to the imaginary unit in the frame of the model of the so-called Gaussian number plane, did it become a legitimate mathematical object, which subsequently assumed an important role in function theory during the 19th and 20th centuries. The complex number plane is a metaphor, because numbers conventionally have nothing to do with geometry. Some, having its origin in mind, assume that the complex number can be interpreted only as a pair of real numbers. Others, following Gauss, consider complex numbers as mathematical objects in their own right.

Gauss begins his "Theory of Biquadratic Residues" in the *Göttingische Gelehrte Anzeige* of 1831 by assuming that to anyone who is not very familiar with the "nature of imaginary quantities" the latter may appear scandalous and unnatural and could lead to an attitude in mathematics that "moves entirely away from intuition. Nothing would be more unfounded than such a view", Gauss writes. As opposed to that view,

> The arithmetic of complex numbers is capable of concrete visualization. [...] Just as the absolute whole numbers are represented by a series of points distributed on a straight line at equal distances, [...] the representation of complex numbers requires but the addition that series be considered as being situated in a determinate unbounded plane, and that parallel to it an unlimited number of similar series is assumed at equal distances from one another, resulting in our having before us, instead of a series of points, a system of points which can be aligned in a twofold way into series of series. [...] In this representation, the execution of the arithmetical operations becomes capable, with regard to the complex quantities, of a representation that leaves nothing to be desired. Thereby the true metaphysics of imaginary numbers is placed in a new light.

4. PEIRCE'S THEORY OF MEANING

In this section, I would like to outline Peirce's theory of meaning, hoping to shed further light on the complementarity that characterises the notion of meaning and, in particular on the question of mathematical existence. Peirce's theory of meaning exhibits the problem, which Thom had expressed so emphatically in a very distinct manner. On the one hand, pure mathematics is for Peirce, as it was for Kant, not knowledge of something, but is concerned with the general possibilities of knowledge. Mathematical meanings lack a definitely referential pole.

Peirce considers pure mathematics as the basis of phenomenology and believes phenomenology—"a science that does not draw any distinction of good and bad in any sense whatever, but just contemplates phenomena as they are, simply opens its eyes and describes what it sees; not what it sees in the real as distinguished from figment—not regarding any such dichotomy—but simply describing the object, as a phenomenon, and stating what it finds in all phenomena alike" (1958 V, par. 37)—to be the basis of all other sciences.

Perhaps you will ask me, Peirce continues, "whether it is possible to conceive of a science which should not aim to declare that something is positively or categorically true. I reply that it is not only possible to conceive of such a science, but that such science exists and flourishes, and Phenomenology, which does not depend upon any other positive science, nevertheless must, if it is to be properly grounded, be made to depend upon the Conditional or Hypothetical Science of Pure

Mathematics, whose only aim is to discover not how things actually are, but how they might be supposed to be, if not in our universe, then in some other" (1958 V, par. 40).

In mathematics, the translation into a conditional is assumed everywhere. Remember my rephrasing of the statement "x is an odd number" (see previous section). Transforming it into a conditional statement, one might say, for instance, "If x is odd and is divided by 2, there will by definition remain a remainder of 1". From this, one may infer that there is for each odd number x another number n such that $x = (2n + 1)$ holds, and so on. And when we seemingly make a categorical assertion, like "the sum of the interior angles of a triangle equal 180 degrees", its truth is in fact hypothetical and is to be considered relatively to the system of Euclidean axiomatic. The assertion than runs thus: "If Euclid's parallel postulate is valid, then the sum of angles in the triangle is 180 degrees".

Mathematical development seems to proceed essentially by deductive reasoning. Peirce even defined mathematics to be "the science which draws necessary conclusions," adding that it must be defined "subjectively" and not "objectively". He noted "that the essence of mathematics lies in its making pure hypotheses, and in the character of the hypotheses which it makes. [...] Hence to say that mathematics busies itself in drawing necessary conclusions, and to say that it busies itself with hypotheses, are two statements which the logician perceives to come to the same thing" (1958 III, par. 558).

Nevertheless one must admit that "making pure hypotheses" and "drawing necessary conclusions" from these seem to be quite different and even complementary processes. Formal axiomatic mathematics is a closed game with no objectivity involved, apart from the constraints of logic. "Making pure hypotheses" by means of intuition and abductive reasoning, in contrast, is confronted with reality as something completely outside of any determination or description. Intuition is not guided by rules.

Now mathematics can only be that "phenomenological science", Peirce wants it to be, if it is situated beyond dichotomies like that of logic vs. intuition. And this Peirce wants to accomplish by means of his semiotics. All necessary reasoning, Peirce declares, "is mathematical reasoning. Now mathematical reasoning is diagrammatic. This is as true of algebra as of geometry" (1958 V, par. 148). He then presents a geometrical example of diagrammatic reasoning and continues:

> In any case, either in the new diagram or else, and more usually, in passing from one diagram to the other, the interpreter of the argumentation will be supposed to see something, which will present this little difficulty for the theory of vision, that it is of a *general* nature. [...] If you admit the principle that logic stops where self-control stops, you will find yourself obliged to admit that a perceptual fact, a logical origin, may involve generality. This can be shown for ordinary generality. But if you have already convinced yourself that continuity is generality, it will be somewhat easier to show that a perceptual fact may involve continuity than that it can involve non-relative generality.

> [...] Generality, Thirdness, pours in upon us in our very perceptual judgments, and all reasoning, so far as it depends on necessary reasoning, that is to say, mathematical reasoning, turns upon the perception of generality and continuity at every step. (1958 V, par. 148–150, my italics).

Mediation belongs, as was said, to Peirce' category of Thirdness. Mathematical deduction thus represents Thirdness; it mediates between the Firstness of Abduction and the Secondness of inductive verification. Deduction is not formally deterministic, at least, not in the more complex cases. According to Peirce, "deduction is really a matter of perception and of experimentation, just as induction and hypothetic inference are; only, the perception and experimentation are concerned with imaginary objects instead of with real ones. The operations of perception and of experimentation are subject to error, and therefore it is only in a Pickwickian sense that mathematical reasoning can be said to be perfectly certain" (1958 VI, par. 595).

Thus knowledge and meaning must be considered from a developmental or genetic point of view. Mathematical reasoning consists in seeing a relationship between two different diagrams, or rather consists in the translation of one diagram into another one—Peirce speaks here of kinds of experiments conducted on the initial diagram—, which is an interpretant of the first one. Both diagrams are connected by a sort of law, or rather are considered to be token of one and the same type, which is their common sense (Sinn) or meaning and which is an expression of the continuity of the mind. This type is the general, entering into our perceptual judgments. Perceptual judgments are fallible and must be refined and corrected by drawing conclusions from them and by verifying these conclusions by means of perceptions again. The beginning and the outcome of any reasoning must be something perceivable. The objectivity of mathematics is to be seen in the objectivity of its laws or meanings, which are, however, not just the laws of formal logic, but rather the laws of Semiosis, that is, mathematical perceptions does not refer to empirical objects but rather to signs and models.

Mathematical deductions thus are not so different from empirical inferences as both employ intuitive or perceptual judgments. And these judgments always have a quasi-probabilistic character, as they operate vis a vis a continuous and open reality. This does not mean that such inferences are completely at random. Peirce's own term is "agapism". He writes:

> I formulate for the reader's convenience the briefest possible definitions of the three conceivable modes of development of thought [...] The tychastic development of thought will consist in slight departures from habitual ideas in different directions indifferently, quite purposeless and quite unconstrained whether by outward circumstances or by force of logic, these new departures being followed by unforeseen results which tend to fix some of them as habits more than others. The anancastic development of thought will consist of new ideas adopted without foreseeing whither they tend, but having a character determined by causes either external to the mind, such as changed circumstances of life, or internal to the mind as logical developments of ideas already accepted, such as generalizations. The agapastic development of

> thought is the adoption of certain mental tendencies, not altogether heedlessly, as in tychasm, nor quite blindly by the mere force of circumstances or of logic, as in anancasm, but by an immediate attraction for the idea itself, whose nature is divined before the mind possesses it, by the power of sympathy, that is, by virtue of the continuity of mind (1958 VI, par. 307).

This continuity of mind could also be called continuity of representation, or mediation, or the process of meaning. "For every symbol is a living thing, in a very strict sense that is no mere figure of speech. The body of the symbol changes slowly, but its meaning inevitably grows, incorporates new elements and throws off old ones" (1958 II, par. 222). Peirce general term for this is *Thirdness*. Now Thirdness is nothing but mediation, "which reaches its fullness in Representation. Thirdness, as I use the term, is only a synonym for Representation", says Peirce (1958 V, par. 105). A sign is a sign of something to somebody, and as such it is a mediation between the subjective and objective.

"Now it is proper to say that a general principle that is operative in the real world is of the essential nature of a Representation and of a Symbol because its modus operandi is the same as that by which words produce physical effects" (1958 V, par. 105). These effects are the effects on the addressee of the sign, that is, on its interpretant. Let me note at this point that conceiving of meaning and cognition in semiotic terms makes them collective entities to begin with. "The semiotic theory of Peirce is an attempt to explain the cognitive process of acquiring scientific knowledge as a pattern of communicative activity in which the dialogue partners are, indifferently, members of a community or sequential states of single person's mind" (Parmentier, 1994, p. 3), as they appear in terms of diagrams or representations.

A sign or *representamen,* Peirce defines as follows:

> [It] is something which stands to somebody for something in some respect or capacity. It addresses somebody, that is, creates in the mind of that person an equivalent sign, or perhaps a more developed sign. That sign which it creates I call the interpretant of the first sign. The sign stands for something, its object. It stands for that object, not in all respects, but in reference to a sort of idea, which I have sometimes called the ground of the representamen. 'Idea' is here to be understood in a sort of Platonic sense, very familiar in everyday talk. (1958 II, par. 228)

The interpretant is that in which the sign results; it is a sort of translation, and the ground is a perspective on the object. It seems here that Peirce considers the interpretant as the meaning of the sign. But Peirce gives further definitions of sign and of meaning also:

> A sign stands for something to the idea which it produces, or modifies. Or, it is a vehicle conveying into the mind something from without. That for which it stands is called its object; that which it conveys, its meaning; and the idea to which it gives rise, its interpretant. (1958 I, par. 339).

Here the meaning is the referent or object of the sign, it is the "immediate Object, which is the Object as the Sign itself represents it, and whose Being is thus dependent upon the Representation of it in the Sign" (1958 IV, par. 536). There is, however, no absolute distinction between object and interpretant because both are signs themselves.

> The object of representation can be nothing but a representation of which the first representation is the interpretant. But an endless series of representations, each representing the one behind it, may be conceived to have an absolute object at its limit. The meaning of a representation can be nothing but a representation. In fact, it is nothing but the representation itself conceived as stripped of irrelevant clothing. But this clothing never can be completely stripped off; it is only changed for something more diaphanous. So there is an infinite regression here. Finally, the interpretant is nothing but another representation to which the torch of truth is handed along; and as representation, it has its interpretant again. Lo, another infinite series. (1958 I, par. 339).

The concept of meaning is thus dualistically constituted from the very beginning, consisting of the object of the sign in question or rather a perspective on it, that which Peirce had called the "ground", on the one hand, and of the series of its interpretants, on the other. A sign determines its interpreter by producing in this person an effect, an interpretant, which stands to the object in the same sort of relation as that in which the original sign itself stands.

There arises therefore the question as to what is it that unites these endless series into a whole? Meaning would be the relation that links a sign to the next sign (i.e., its interpretant). A sign functions by making relations effective; it functions like a natural law, for instance. This law is what Peirce called the "ground".

And it should also be called the meaning of the sign. The notion of meaning has, as was said already, a dual structure, consisting of the effect or interpretant, jointly with the law that determines the "interpreting" representation as a semiotic process.

For the proper significate outcome of a sign, I propose, writes Peirce, "the name, the interpretant of the sign. [...] it is all that is explicit in the sign itself apart from its context and circumstances of utterance" (1958 V, par. 473). The interpretant corresponds to what I have identified as the intension of a term, whereas the object of a sign corresponds to what I have called the objective component of meaning. What is important here is to notice that the interpretant is deliberately *not* described as being necessarily the interpreter or an idea in the mind of some interpreter, but is rather a semiotic action producing another sign. The "idea" or ground of a sign, the law or the relation between object and interpretant that the sign turns effective, does not exist apart from the sign itself.

A consequence of Peirce's phenomenological view of knowing is that the developmental processes of knowledge and meaning in mathematics and in the empirical sciences bear some resemblance. When Galileo let balls roll down an inclined plane with constant velocity, he was not interested in the event as such but rather searched for a general law governing these phenomena and he formulated this law also as a conditional statement. Almeder (1983) in a very thoughtful paper on Peirce's theory of meaning writes:

> The meaning of any given expression is obtained by translating that expression into a set of conditional statements, the antecedents of which prescribe certain observable phenomena which should and would occur as the result of performing those operations of the proposition were true. For example, consider the expression "hard". The expression means "not scratchable by many other substances"; but that is also equivalent to a set of conditionals, the expression "not scratchable by many other substances" means more properly "If you were to take some object which is said to be hard, and if you were to search it with many substances, then it would not be scratched." In short, for Peirce, the meaning of what he calls "intellectual concept" or proposition is simply the conditions of its verification (p. 329).

What are these conditions, however? If they consisted just in certain descriptions of the situation, this cannot be the whole truth, as we would otherwise be led to ever new conditional statements. Take the example of mathematical proof. Every proof is faced with the request that one proves that the proof is correct. And the proof of the correctness of the proof again meets the same requirement, and also the proof of the correctness of the correctness of the proof, and so on ad infinitum. Lewis Carroll, in his famous little piece "What the Tortoise Said to Achilles" (see Hofstadter, 1979, pp. 43–45), beautifully illustrated the dilemma. In order to escape from an infinite regress, one is led to assume that any argument has to also contain a compulsory element from which there is no reasonable escape. This cannot be done by rational argument alone. The addressee of the argument of the proof must be in a condition where he can no longer avoid *seeing* by himself that the situation is such and such.

At this point the importance of perception or intuition enters once more. In a late manuscript Peirce has described this importance thus:

> I do not think it is possible fully to comprehend the problem of the merits of pragmatism without recognizing these three truths:
>
> 1. that there are no conceptions which are not given to us in perceptual judgments, so that we may say that all our ideas are perceptual ideas. This sounds like sensationalism but in order to maintain this position it is nesessary to recognize,
> 2. that perceptual judgments contain elements of generality; so that Thirdness is directly perceived; and finally I think it of great importance to recognize,
> 3. that the Abductive faculty, whereby we divine the secrets of nature is, as we may say, a shading off, a gradation of that which in its highest perfection we call perception. (1967, p. 316).

Peirce explains the second point, which is of particular importance with respect to mathematical diagrammatic reasoning, in his (6th) Lecture on Pragmatism of 1903 as follows:

> In saying that perceptual judgments involve general elements I certainly never intended to be understood as enunciating any proposition in psychology. [...] All that I can mean by a perceptual judgment is a judgment absolutely forced upon my acceptance, and that by a process which I am utterly unable to control and consequently am unable to criticize. Nor can I pretend to absolute certainty about any matter of fact. If with the closest scrutiny I am able to

> give, a judgment appears to have the characters I have described, I must reckon it among perceptual judgments until I am better advised. Now consider the judgment that one event C appears to be subsequent to another event A. Certainly, I may have inferred this; because I may have remarked that C was subsequent to a third event B which was itself subsequent to A. But then these premises are judgments of the same description. It does not seem possible that I can have performed an infinite series of acts of criticism each of which must require a distinct effort. The case is quite different from that of Achilles and the tortoise because Achilles does not require to make an infinite series of distinct efforts. It therefore appears that I must have made some judgment that one event appeared to be subsequent to another without that judgment having been inferred from any premise [i.e.] without any controlled and criticized action of reasoning. If this be so, it is a perceptual judgment in the only sense that the logician can recognize. But from that proposition that one event, Z, is subsequent to another event, J, I can at once deduce by necessary reasoning a universal proposition. Namely, the definition of the relation of apparent subsequence is well known, or sufficiently so for our purpose. [...] It easily follows that whatever is subsequent to C is subsequent to anything, A, to which C is subsequent—which is a universal proposition. Thus my assertion at the end of the last lecture appears to be most amply justified. Thirdness pours in upon us through every avenue of sense (1958 V, par. 157)

5. PEIRCE'S THEORY OF MEANING CONTINUED

In his groundbreaking papers on the theory of information, Shannon defined a quantity that he called *amount of information,* which is essentially measured in terms of statistical unexpectedness. Since unexpectedness of a message may even be contrary to its meaningfulness, Shannon insisted that the concept of "meaning" was outside the scope of information theory. This assertion has given rise to the unfortunate consequence, particularly in the debate about computer thought, of a seeming opposition between the mechanical and the spiritual, to a mind-body dualism. Proposals to ignore this dualism by defining the meaning of a message simply as the behavior pattern it produces in the receiver, will, for various reasons, not do. Information theory nevertheless tried to overcome this unhappy dualism by a functional approach to the problem of meaning, asking, for instance, what difference it makes when somebody receives and understands the meaning of a message. The effect of a message, MacKay (1969), answering this question, says

> is not necessarily what you do—as some behaviorists have suggested—but what you would be ready to do, if given (relevant) circumstances arose. [...] It is not your behavior, but rather your state of conditional readiness for behavior, which betokens the meaning (to you) of the message you heard. (p. 22)

This disposition or conditioned readiness is what Peirce called a habit, and he conceived of the meaning of a symbol in terms of habits. Whereas an act depends on certain specific circumstances, a habit represents an indefinite number of acts, corresponding to an indefinite number or conditions or circumstances.

> Le caractère d'une habitude dépend de la façon dont elle peut nous faire agir non pas seulement dans telle circonstance probable, mais dans toute circonstance possible, si improbable qu'elle puisse être. Ce qu'est une habitude dépend de ces deux points: quand et comment elle fait agir. Pour le premier point: quand? tout stimulant à l'action dérive d'une perception; pour le second point: comment? le but de toute action est d'amener au résultat sensible. Nous atteignons ainsi le tangible et le pratique comme base de toute différence de pensée, si subtile qu'elle puisse être. (1958 V, par. 18)

Habits or meanings thus are general objects or universals that are effective with respect to human behavior. How can they be known, evaluated and if necessary changed or generalised. By means of particular applications, is the seemingly obvious answer. But no law or habit may be reduced and verified by any set of particular applications. Meanings are general ideas or generalisations like natural laws. And like these they cannot be tested or objectively established by any number of applications. And human action also has to be evaluated twice: with respect to its immediate motifs and effects as well as in relation to its generalisability and general relevance.

Peirce's Pragmatic Maxim reflects the problematic. Pragmatism has become known and identified by Peirce's so-called "pragmatic maxim", which should be understood as the cornerstone of a theory of meaning. Its formulation, as Peirce originally stated it in 1878, is as follows:

> Considérer quels sont les effets pratiques que nous pensons pouvoir être produits par l'objet de notre conception. La conception de tous ces effets est la conception complète de l'objet [Consider what effects that might conceivably have practical bearings we conceive the object of our conception to have. Then our conception of these effects is the whole of our conception of the object]. (1958 V, par. 18)

Peirce then explains further:

> To develop the meaning of a thought, it is simply necessary to determine what habits it produces because the meaning of a thing consists simply in the *habits* it implies. The character of a habit depends on the way in which it can make us act [...] in all possible circumstances, however improbable they might be. (1958 V, par. 18)

But what might be understood as the "practical effects" of incommensurables or infinitesimals or natural laws? What could possibly count as a justification in terms of practical consequences for the introduction of these and numerous other "imaginary" theoretical notions? Having the practical consequences of knowledge in mind, Bishop Berkeley, whom Peirce considered his teacher, apodictically stated, "There are no incommensurables, no surds, I say the side of any square may be assigned in numbers". With respect to infinitesimals, he formulated as an axiom: "No reasoning about things whereof we have no idea. Therefore no reasoning about

Infinitesimals". To have an idea for Berkeley meant to have a representation or to be able to make a perceptual judgment.

Thus the pragmatic maxim seems to suggest that pragmatism is but a very crude type of philosophical practicism or utilitarism and that its epistemology is based on a verificationist theory of truth. Peirce himself commented on this problem in 1902 in a contribution to Baldwin's *Dictionary of Philosophy and Psychology:*

> [The pragmatic maxim] might easily be misapplied, so as to sweep away the whole doctrine of incommensurables, and, in fact, the whole Weierstrassian way of regarding the calculus. [...] The doctrine appears to assume that the end of man is action—If it be admitted, on the contrary, that action wants an end, and that end must be something of a general description, then the spirit of the maxim itself, which is that we must look to the upshot of our concepts in order rightly to apprehend them, would direct us towards something different from practical facts, namely, to general ideas, as the true interpreters of our though. (1958 V, par. 3)

This statement is very vague, and it merely indicates that meaning really is a very complex and profound notion. The text by Peirce from Baldwin's *Dictionary* quoted above continues as follows:

> Nevertheless, the maxim has approved itself to the writer, after many years of trial, as of great utility in leading to a relatively high grade of clearness of thought. He would venture to suggest that it should always be put into practice with conscientious thoroughness, but that, when that has been done, and not before, a still higher grade of clearness of thought can be attained by remembering that the only ultimate good which the practical facts to which it directs attention can subserve is to further the development of concrete reasonableness; so that the meaning of the concept does not lie in any individual reactions at all, but in the manner in which those reactions contribute to that development. Indeed, in the article of 1878, above referred to, the writer practiced better than he preached; for he applied the stoical maxim most unstoically, in such a sense as to insist upon the reality of the objects of general ideas in their generality. (1958 V, par. 3)

Therefore meaning conceived of objectively is nothing but a universal force or a law of mind, that is, a habit. In his very first statement of the pragmatic maxim, Peirce had already declared that "the meaning of a thing consists simply in the habits it implies" (1958 V, par. 18). And in a letter to Lady Welby in 1902 he explained, "It appears to me that the essential function of a sign is to render inefficient relations efficient—not to set them into action, but to establish a habit or general rule whereby they will act on occasion" (1958 VIII, par. 332). Thus the final interpretant of a sign is not an action or a thought or a concept or a decision, but rather a habit or a law. Habits, and universals in general, must be explained in genetic terms, as Peirce insisted again and again (see, e.g., 1982 IV, p. 547). Facts and laws, or things and universals have, according to Peirce, an ontological status to be distinguished only relatively. Explanation, for example, is required with respect not only to facts or phenomena, but also to laws (and second-order laws—"laws of laws"—as well, and so forth). The infinite regress involved must be viewed in dynamic terms, not to lead to aporia. Now in order to really acknowledge the constitutive role of the evolutionary character of reality, one has to recognise its probabilistic character. Peirce says that there are two elements in nature: spontaneity and law. Spontaneity is important to Peirce because the heterogeneity and the manifoldness of nature are due to it. "This has not been produced by the operation of law. To prescribe that under given circumstances a fixed result shall occur is to prescribe that the substantive manifoldness of nature shall never be increased." If these two elements of spontaneity and law exist in nature, however, it is clear that what has to be explained are not the facts or the things, but rather the lawfulness. "But to explain a thing is to show it may have been a result of something else. Law, then, ought to be explained as a result of spontaneity" (1967, p. 954). This is going to be done now by means of an example.

5.1 The mouse and the lawn

Let us assume a mouse wishes to cross a meadow, and it finds before its eyes a meadow where all the blades of grass are aligned even more regularly than on the best-trimmed English lawn. The mouse will have to select his own path spontaneously and without a reason for there is no indicator within the lawn's continuity which would help in selecting this course or that. Perception reposes, as we well know, not on light, but rather on differences. At the beginning, there are no differences at all to be found, in this lawn. It is totally homogeneous. As soon as the mouse has once run across, some of its small blades will have been dislocated, however light-footed the mouse may be. And it may be assumed that while the mouse will not necessarily select precisely the same path for a second run across the meadow, it will nevertheless select a similar one. In the course of time, the mouse's traces will become more and more visible, until a well-established mouse-path cuts through the meadow at last. The lawn's continuity has been broken, and the mouse now can determine its course at a glance. The mouse, however, does no longer determine its course at all, but quite to the contrary, it is the established path, which determines the mouse's behaviour now. From the mouse's view, it is a habit to follow this established path. From the path's view, this is a case of a law, i.e. of determining the mouse's movement.

Peirce has described the general rule drawn from this example in another manuscript from 1884 bearing the title "Design and Chance". In this manuscript, Peirce assumes "that all known laws are due to chance and repose upon others far less rigid themselves due to chance and so on in an infinite regress, the further we go back the more indefinite being the nature of the laws, and in this way we see the possibility of an indefinite approximation toward a complete explanation of nature. Chance is indeterminacy, is freedom. But the action o freedom issues in the strictest rule of law" (1982 IV, p. 551). Thus I end with a new description of the meaning of a sign, conceiving it in terms of a habit change.

Peirce calls this change of habit the (ultimate) logical interpretant, and it is the most important means for a person to "exercise more or less control over himself [or herself]" (1958 V, par. 487). The ultimate or final logical interpretant is the meaning of an intellectual concept or sign. It cannot just be a sign or concept itself, because it is not just general but rather has real effects. It also cannot just be feeling or acting, because these lack the necessary generality. It is what Peirce calls a habit.

A habit, interpreted in terms of the ultimate logical interpretant, exhibits a rather strange logical structure, as it is both a "collection-as-many", to borrow some set-theoretical terminology (habits result in actions) and a "collection-as-one" as well (habits or meanings are conditions of actions). A habit simultaneously represents experience of knowledge and of its application, experience of an item of content and of the conditions of its verification. Habits are meanings, and meanings are, as said above, simultaneously both universals and particulars. Consciousness of habit, says Peirce, "is a consciousness at once of the substance of the habit, the special case of application, and the union of the two" (1958 VIII, par. 304). Habits clearly transcend consciousness, although learning conceived of as habit change may on occasion transform all or part of the unconscious into consciousness.

Peirce believed that the world essentially evolves out of a continuum of mere, unspecified, and undifferentiated possibility and chance. It is because all things swim in the continuum of space and time that it is theoretically impossible for us to specify all their properties and hence render our propositions fully determinate with respect to meaning. This second argument convinced Peirce that the mathematics of the continuum would provide a logic for the universe and, in effect, the key that would open the door of his cosmology; but it also furnished the logical foundation for a most distinctive characteristic of his theory of meaning. (Almeder, 1983, p. 332)

Peirce was against the axiom "that real things exist or in other words, what comes to the same thing, that every intelligible question whatever is susceptible in its own nature of receiving a definitive and satisfactory answer" (1982 IV, p. 545; see also 1958 II, par. 113), or again in still other words "that every event has a cause". Stated in semiotic terms, this argument comes down to the following two claims: First, for something to have a meaning, it must be related to something else. Nothing can just mean itself, although that has many times been exactly the goal of logical constructions of so-called ideal languages. But, secondly, meaning cannot be related to anything arbitrarily chosen. There are always constraints to an act of interpretation. These constraints cannot, however, be specified once and for all.

Meanings thus are continua. As Sandra Rosenthal (1983) explains:

> Thus, while the ontological dimensions of habit lead to the expression of the validity or appropriateness of meanings in terms of the ongoing conduct of the biological organism emerged in a natural world, the epistemic dimensions of habit lead to their expression in terms of the phenomenological description of the appearance of what is meant. It is the epistemic dimension of meaning in terms of habit, which provides, further, the source of a sense of the concrete unity of objectivity as more than a collection of appearances. Just as a continuum may generate an unlimited number of cuts within itself, so a disposition as a rule of organization and generation contains within itself an unlimited number of possibilities of specific acts to be generated. As Peirce

states, 'a true continuum is something whose possibilities of determination no multitude of individuals can exhaust,' while a habit or general idea is a living feeling, infinitesimal in duration and immediately present, but still embracing innumerable parts. (p. 316)

REFERENCES

Almeder, R. (1983). Peirce on meaning. In E. Freeman (Ed.), *The relevance of Charles Peirce* (pp. 328–347). La Salle (USA): Open Court.

Beth, E. W. & Piaget, J. (1966). *Mathematical epistemology and psychology*. Dordrecht: Reidel.

Boole, G. (1958). *The laws of thought*. New York: Dover. (First published 1853.)

Bruner, J. (1969). *The process of education*. Cambridge (USA): Harvard University Press.

Cassirer, E. (1962). *An essay on man*. New Haven (USA): Yale University Press.

Castonguay, C. (1972). *Meaning and existence in mathematics*. Vienna: Springer.

Courant, R. & Robbins, H. (1941). *What is mathematics?* London: Oxford University Press.

De Gandt, F. (1995). Force and geometry in Newton's Principia. Princeton: Princeton University Press.

Eisele, C. (1976) (Ed.). The new elements of mathematics by Charles S. Peirce, Vol. I-IV. Paris: Mouton.

Frege, G. (1892). Über Sinn und Bedeutung. *Zeitschrift für Philosophie und philosophische Kritik, 100*, 25-50. (Translated as "On Sense and Reference" in P. Geach & M. Black (Eds. and Trans.) (1980), *Translations from the Philosophical Writings of Gottlob Frege*. Oxford: Blackwell)

Frege, G. (1980). Brief an Hilbert vom 27.12.1899 [Letter to Hilbert on 27 December 1899]. In G. Gabriel, F. Kambartel & C. Thiel (Eds.), *Gottlob Freges Briefwechsel mit D. Hilbert, E. Husserl, B. Russell, sowie ausgewählte Einzelbriefe Freges* (pp. 6–10). Hamburg: Meiner. (Original work published 1899)

Grayling, A. C. (1977). *An introduction to philosophical logic*. Oxford: Blackwell.

Hacking, I. (1980). Descartes and Leibniz: Proof and eternal truths. In S. Gaukroger (Ed.), *Descartes: Philosophy, mathematics and physics*. Sussex: Harvester.

Hilbert, D. (1992). *Natur und mathematisches Erkennen: Vorlesungen, gehalten 1919-1920*. Göttingen: Birkhäuser.

Hintikka, J. (1997). *Lingua universalis vs. calculus ratiocinator*. Kluwer: Dordrecht.

Hofstadter, D. (1979). *Gödel, Escher, Bach: An eternal golden braid*. Hassocks (UK): Harvester.

Kant, I. (1933). *Critique of Pure Reason* (N. Kemp Smith, Trans.). London: MacMillan. (First German edition 1791).

Lovejoy, A. O. (1936). *The great chain of being*. Cambridge (USA): Harvard University Press.

MacKay, D. M. (1969). *Information, mechanism and meaning*. Cambridge (USA): M.I.T. Press.

McGinn, C. (2000). *Logical properties*. Oxford: Claredon.

Nahin, P. (1998). *The story of* $\sqrt{-1}$. Princeton: Princeton University Press.

Otte, M. (1990). Arithmetic and geometry: Some remarks on the concept of complementarity. *Studies in Philosophy and Education, 10*, 37–62.

Otte, M. (1991). Style as a historical category. *Science in Context, 4*, 233–264.

Otte, M. (1994). *Das Formale, das Soziale und das Subjektive: Eine Einführung in die Philosophie und Didaktik der Mathematik* [The formal, the social and the subjective: An introduction to the philosophy and didactics of mathematics]. Frankfurt: Suhrkamp.

Otte, M. (2002). *Proof, analysis and continuity*. Bielefeld: Institut für Didaktik der Mathematik.

Otte, M., Steinbring, H. & Stowasser, R. (1977). *Mathematik die uns angeht* [Mathematics that concerns us]. Gütersloh: Bertelsmann.

Parmentier, R. (1994). *Signs in society*. Bloomington (USA): Indiana University Press.

Peirce, C. S. (1958). Collected papers of Charles Sanders Peirce, Volumes I-VI (C. Hartshorne & P. Weiß, Eds.), Volumes VII-VIII (A. W. Burks, Ed.). Cambridge: Harvard University Press.

Peirce, C. S. (1967). Manuscript, according to Richard S. Robin, Annotated Catalogue of the Papers of Charles S. Peirce. Amherst (USA): The University of Massachusetts Press.

Peirce, C. S. (1982). Writings of Charles S. Peirce. A Chronological Edition, Vol. I-V. Bloomington (USA): Indiana University Press.

Piaget, J. (1981). *Le possible et le necessaire* [The possible and the necessary]. Paris: Presses Universitaires de France.
Quine, W. V. (1969). *Ontological relativity.* New York: Columbia University Press.
Rosenthal, S. B. (1983). Meaning as habit: Some systematic implications of Peirce's pragmatism. In E. Freeman (Ed.), *The relevance of Charles Peirce* (pp. 312–327). La Salle (USA): Open Court.
Russell, B. (1970). *Introduction to mathematical philosophy.* London: Allen & Unwin.
Russell, B. (2000). *History of Western Philosophy.* London: Routledge.
Thom, R. (1973). Modern mathematics: Does it exist? In A. G. Howson (Ed.), *Developments in mathematical education.* Cambridge: Cambridge University Press.
Thun, F. & Götz, H. (1976). *Mathematik verstaändlich erklären.* München: Urban Schwarzenberg.
Toerner, G. (2003). Schwierigkeiten bei der mentalen Repräsentation von irrationalen Zahlen. In *Beiträge zum Mathematikunterricht* (pp. 629–632). Berlin: Franzbecker.

MIX
Papier aus verantwortungsvollen Quellen
Paper from responsible sources
FSC® C105338

If you have any concerns about our products,
you can contact us on
ProductSafety@springernature.com

In case Publisher is established outside the EU,
the EU authorized representative is:
Springer Nature Customer Service Center GmbH
Europaplatz 3, 69115 Heidelberg, Germany

Printed by Libri Plureos GmbH
in Hamburg, Germany